The Oldest Rocks on Earth

The Oldest Rocks on Earth

A Search for the Origins of Our World

Simon Lamb

Columbia University Press

New York

Columbia University Press
Publishers Since 1893
New York   Chichester, West Sussex
cup.columbia.edu
Copyright © 2026 Columbia University Press

Cataloging-in-Publication Data is available from the Library of Congress.
ISBN 9780231222228 (hardback)
ISBN 9780231222235 (trade paperback)
ISBN 9780231564113 (epub)
LCCN 2025022000

Cover design: Julia Kushnirsky
Cover image: View of the Makhonjwa Mountains in northwest Eswatini
(formerly Swaziland). Shutterstock.

GPSR Authorized Representative: Easy Access System Europe, Mustamäe tee 50,
10621 Tallinn, Estonia, gpsr.requests@easproject.com

Contents

Illustrations

Preface

On Christmas Eve 1968, during the Apollo 8 space mission, William Anders snapped what has been described as the most influential environmental photograph ever taken. He was in the lunar orbiter, and as it circled the moon, he managed to capture Earth appearing above the moon's horizon, becoming one of the first humans to witness an earthrise. In the photograph, Earth appears as a vibrant blue ball, partly obscured by Earth's own shadow and streaked with swirly white clouds. We see mainly the Atlantic Ocean, with parts of Africa and South America visible as orange and green landmasses. This image of Earth is all the more striking because it is in stark contrast to the lunar surface that forms a gray and barren foreground pulverized by meteorite bombardment. With his eye pressed against the viewfinder of the camera, it may have crossed Anders's mind that the very things giving Earth its distinctive blue appearance—an abundance of liquid water, a relatively thick atmosphere, and a world constantly reshaped by geological activity—were the reasons he existed at all.

The experience of looking back at Earth from space has proved to be deeply emotional for almost all astronauts, making them acutely aware of how small and precious our terrestrial home is, truly a haven for life surrounded by the nothingness of space. As one astronaut remarked, on the moon you could blot out our living world with your thumb held up in front of you. In this context, our planet seems so fragile, yet, paradoxically perhaps, we also know that it has been around for a very long time. Cosmologists have measured time itself, starting with the Big Bang at the creation of the universe, to be about 14 billion years old. It is hard to imagine such an immense duration. Earth, occupying its place in the solar system, has existed for about a third of this—about 4.6 billion years. We, as *Homo sapiens*, have only been here for a mere blink in this vast perspective, less than 0.002 percent of the planet's history. Given the great age of Earth, can we be sure that it was always the welcoming place for life we know today? If we ventured far back into the past, would we reach a time when it was fundamentally different, and no longer had the essential features of the blue planet in William Anders's *Earthrise* photograph? Put another way, when, and importantly why, did the place we call our home become the blue planet? And what does this tell us about its long-term robustness? It seems to me that with the specter of global warming and other environmental crises, answering these questions has assumed a special significance, because it will give us a new perspective on how Earth might respond to these threats.

It is geologists who address questions like these through their study of the planet's bedrock, in the process reaching into its deep past. I am a geologist myself, and I am fascinated by this idea of time travel. I also have many more questions I want to ask. So, let's indulge ourselves in a flight of fancy and imagine another type of time traveler, in this case extra-terrestrial alien beings who visited our part of the solar system billions of years ago, during the early years of our planet's story. In addition, let's imagine that these travelers perceived the external world in the same way that we do, and they deployed some sort of remotely operated vehicle to explore the surface of Earth—like the NASA rovers on Mars—bristling with sensors and beaming back pictures and data. What was the terrestrial surface like in these images? Did the images show oceans, mountain ranges, rivers? Video recordings may have even captured a volcanic eruption, but what sort of eruption, and was the surface also shaped by great earthquakes, drifting continents and plate tectonics that have all played profound roles in shaping our world today? And would humans have been able to survive here, or was it an alien and perhaps dangerous place? Was there any life at all?

Extraordinary as it may seem, these are not impossible questions to answer. This book is about my own search for answers.

In remote corners of the world, geologists have found relics of the surface of our early planet, forming Earth's oldest rocks. Most are just shadows of their former selves, strongly altered over time. A few, however, are almost pristine and surprisingly extensive, and some of the oldest are in Africa. How they survived the mind-boggling spans of geological time is a marvel of the natural world. But survive they did, and like a written archive that provides a detailed record of a long-lost civilization, they contain an enormous amount of information about early Earth. The rock archive, however, is not easy to read, and so geologists have been forced to be very ingenious to understand its full meaning. Early on in my career as a geologist, I had the chance to join this scientific endeavor, looking at the bedrock of the Makhonjwa Mountains in the small southern African kingdom of Eswatini (formerly called Swaziland) and the adjacent part of South Africa. There are rocks here that reach back over 3.5 billion (3,500 million) years, to a time which was less than a quarter of the way through Earth's long history.

Whenever I think back to the beginnings of this project, I am amazed at what I launched myself into, the people I met, and the risks I took—all these have shaped the rest of my life. Many of the puzzling observations from this early research have stayed with me ever since, and it has only been after spending many years studying the youthful and earthquake-prone mountain ranges in New Zealand and South America that I feel I know enough to make sense of the rocks in the Makhonjwa Mountains. I have come to realize that these samples of Earth's oldest rocks contain the critical evidence for how the foundations of the world we live in were first built, as well as the origins of the forces that have shaped Earth's subsequent history. Life was thriving too, colonizing much of the planet's surface and poised to take an important role in controlling this history. With this rich image of these early years, we can now make use of rarer clues in even older rocks to find the true beginnings of Earth's story, journeying all the way back to the year zero at the birth of the solar system. We will then be able to answer not only the question *when* Earth became the living blue planet, but also *why*, uniquely amongst its planetary neighbors.

Each chapter of the book contains two intertwined stories. One is a scientific story about reading the rocks and working out what they tell us about early Earth, uncovering the landscapes or seascapes of the planet at this time, and how these "scapes" were created. There is also a personal story about me and scientific colleagues on their quests for answers. I hope that by the end of the book, the early years of our planet will seem like a well visited place, and the scientists who took you there almost old friends!

My geological endeavors were just one small part of a much bigger and global scientific enterprise involving geologists, geophysicists, geochemists, geochronologists, paleontologists, and many more. Amongst these scientists, there was one who clearly stood out for me—Maarten de Wit—and you will hear much more about him throughout this book. He introduced me to the geological wonders of the Makhonjwa Mountains and kept me on my toes with so many thought-provoking ideas about early Earth. Maarten died in April 2020, and his passing away made me realize how much I was going to miss talking to him about rocks. He was one of Africa's most distinguished earth scientists and held the Chair of Earth Stewardship Science at Nelson Mandela University in South Africa. In my mind, Maarten was one of those rare individuals who really made a difference, enriching the lives of so many others. It is this that has motivated me to write this book before I forget what went before, and I dedicate it to his memory.

Many others have supported me over the years in my geological research. My parents encouraged me to be adventurous and were always there to pick up the pieces when things—as they often did—went wrong. My thesis supervisor Alan Smith listened patiently to my wild ideas and kept me on the right track with his wise advice. My thesis research was supported by the UK Natural Environment Research Council, the British Geological Survey, the Harkness Scholarship from the Department of Earth Sciences, Cambridge, and a Research Scholarship and Ver Herbert de Lancy Prize from Trinity College, Cambridge. I would like to thank the friends, colleagues and people I came in contact with in Eswatini and South Africa, who opened up their homes to me or helped me in many critical ways during my long field seasons—in particular, the Barton and Bateson families, Fergs de Wit, Anna Ekberg, Margy Jeffery, Isabelle Paris, Ian Stanistreet, and those Peace Corps volunteers and aid workers I got to know, and also the geologists in what was then the Swaziland Geological Survey; the Swaziland National Trust Commission park wardens; the pragmatic police officer who made it possible for me to use a motorbike; and the mechanic at the local garage in Mbabane who repeatedly fixed my Land Rover. Lou Guarino was a patient, kind and forgiving field assistant during my 1981 field season. Dick Walcott, Professor of Geology at Victoria University of Wellington, was instrumental in the award of a three-year University Grants Committee postdoctoral research fellowship in New Zealand; this opened my eyes to the many similarities between the very young geology in New Zealand and the ancient rocks in Swaziland. Subsequently, I was supported by a one-year BP research fellowship, and then a six-year UK Royal Society University Research Fellowship to work in the Andes Mountains of South America. I continued my research as a lecturer in geology at the University of Oxford before coming to Victoria University of Wellington.

I am indebted to my colleagues who heroically waded through earlier versions of the manuscript, providing me with essential feedback and encouragement, particularly Tim Stern who read the manuscript more than once, but also Rodney Grapes, Cornel de Ronde, Harald Furnes, Laurence Robb, and Tony Watts—beer is on me the next time we meet! My daughter Clara made the beautiful landscape sketches of New Zealand and early Earth, all of which greatly enrich the book. Finally, without the enthusiasm and support of my editor Miranda Martin, and her production team, this book would never have been published. In particular, I would like to thank: Ben Kolstad and Kathryn Jorge for sensitively and tirelessly managing the publication process with the assistance of Miriah Ralston; Chang Jae Lee for their wonderful book design; Marcella Munson for carefully copyediting the manuscript; and Max McMaster for the comprehensive index.

• • •

Please note that I have used "Ma" and "Ga" in some figures as a compact way to deal with the huge spans of geological time, where "Ma" denotes ages in millions of years and "Ga" in billions of years.

Act I

1 | Some Beginnings

Becoming a geologist: rocks, deep time, tectonic plates, and a new project in Africa

It was clear to me at school that I was going to be a scientist. To begin with, I was determined to be an astronomer, then an archaeologist, and finally a zoologist. So I went to Cambridge University to study biology, but after one year spent grappling with the endlessly complicated biochemical reactions inside a single living cell, I started to suffer from claustrophobia. A friend of my father was a geologist at the university. This had alerted me to geology, and I soon discovered that the wide scope of the subject—described by the geology department as the study of the planet we live on—was the perfect antidote to cellular biology. And fortunately, I had the opportunity to take it up as an extra course. The department had a museum, known as the Sedgewick Museum, which was full of fossils and samples of rock from all over the world. The museum's display cabinets were a catalogue of strange places and names, and no place seemed too difficult for a geologist: the high peaks of the

Himalayas, the Andes, the hot interiors of Africa and India, and the polar regions of Greenland, Spitsbergen and Antarctica. Geologists had been virtually everywhere, sending the best fossil and rock specimens back to the geology department. I wanted to be fully involved in this scientific endeavor. So, in my second year at university, I switched to geology. And, as a second-year undergraduate, I even managed to add to the Sedgewick collection with a rare fossil of my own.

By the end of my third year, I was certain that I wanted to continue in geology by studying for a doctorate. Possible projects were advertised on the department's notice board. Two caught my eye. One involved working in the Arctic regions, on the island of Spitsbergen. In the winter, Spitsbergen was a snow- and icebound place. But in the summer, when the snow had melted, a rocky landscape emerged, inviting examination by geologists. These rocks held the secrets of the distant geological origins of northern Europe and the birth of the Atlantic and Arctic Oceans, subjects of great interest to both academics and mining companies. Every summer, there was money enough to pay for a group of scientists to sail north from Denmark, unload their expedition equipment on the barren coastline, and set up camp for several months. They scrambled over the bedrock, recording, measuring and photographing myriad details (armed, I was told, with guns to shoot aggressive polar bears). And now a student was needed to carry on with this work for a doctoral dissertation. This could be the project for me.

But there was another project. Curiously, I did not at first concern myself too much about the details, although it involved studying some very ancient rocks, which I was told were relics of early Earth. What caught my imagination was where these rocks were located: in the small southern African kingdom of Swaziland (now Eswatini). When I was a child, my parents had lived and worked in West Africa, travelling extensively through the southern fringes of the Sahara. I was at boarding school in England at the time, but I longed to be with them on these journeys. Now, I thought, I have the chance to go on my own expedition. As I naively imagined it, the hot grasslands would be dotted with low rocky hills. I saw myself at sunset, camped out near one of these hills, relaxing after a hard day examining the rocks. In the warm orangey glow of this image, Spitsbergen seemed white and chilly. My mind was set on Africa.

SOLID STUFF

For those who do not know much about Africa, I suspect there is a tendency to view it, as I did before starting out on my project, in rather simplistic ways, despite the fact that Africa is home to roughly 1.3 billion people living in over fifty countries with widely different languages, cultures, and histories. If you read too many old-fashioned adventure books, perhaps it is a land of tribal wars and fierce animals—a land of lions and tigers (but tigers don't live in Africa)? Or if you watch wildlife documentaries, perhaps wide expanses of grassland, grazed by limitless herds of buffalo, antelope, giraffes, and elephants? Or if you are an armchair explorer, perhaps a continent with thick impenetrable tropical jungle, long rivers, and the Sahara Desert? Or if you study nineteenth- and twentieth-century history, a place colonized and exploited by European countries; and in the south, South Africa, once the heartland of the cruel Apartheid regime?

As a geologist, I now see Africa through a very different lens. It is a continent, forming a large continuous area of dry land. This puts it in the elite one-third of the surface of our planet—the remaining two-thirds, of course, are ocean. But a continent is not just a lack of a covering of water. From the point of view of a geologist, it is the underlying bedrock that is important. The bedrock of the continents has proved to be particularly rich pickings, and Africa has some of the richest. Its bedrock contains vast mineral wealth, such as diamonds, gold, copper, iron, uranium, and many other metals these constitute a significant proportion of the planet's entire mineral resources. But there is vast scientific wealth too, because, as we shall see, this bedrock is full of information about the history of our planet.

My project in southern Africa was part of a scientific enterprise to investigate the very early years of Earth. I should say that my own motivation in all this, once I had got beyond the excitement of working in Africa, was very simple: I had been long fascinated by what it would be like to travel back in time, and I now realized that the ultimate journey into the past was one that reached the other end of our planet's history, with the aim of exploring its very early surface. To understand why this might also be scientifically important, and how it is even possible, I need to introduce you to ideas about our planet that may be unfamiliar unless you are a geologist

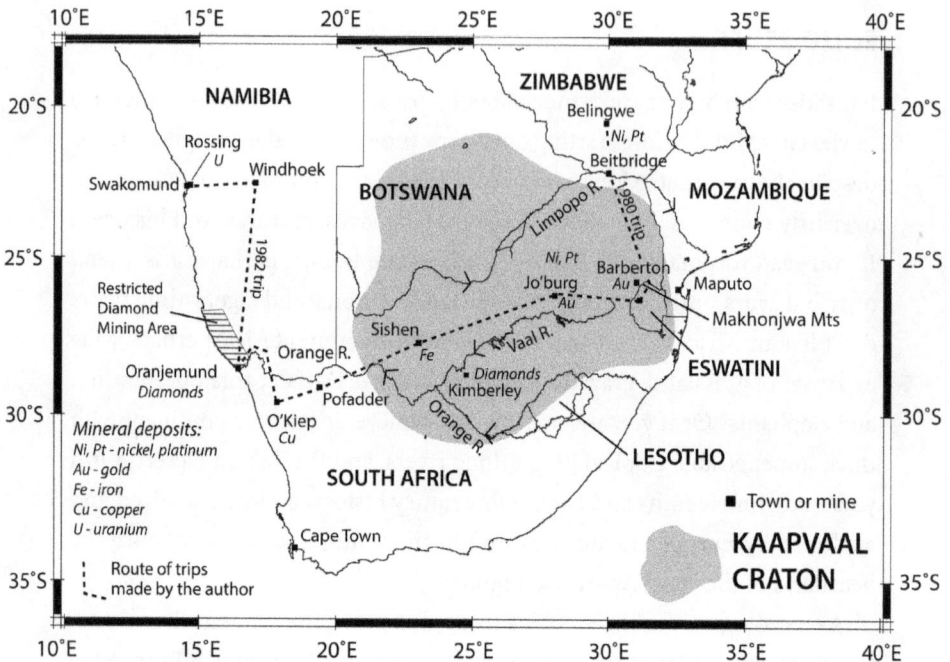

FIGURE 1.1 Southern Africa comprises many countries and spans a region over 1,500 kilometers (about 900 miles) across. Major rivers, such as the Vaal, Orange, and Limpopo, flow through the region, in some areas defining international borders. The region is underlain by some of the oldest rocks on Earth, including the ancient Kaapvaal Craton in the south, and is rich in mineral deposits–I visited several mines during my research in 1982, and in 1980, I travelled to Zimbabwe to study the ancient rocks near Belingwe. I was based in Eswatini, in the Makhonjwa Mountains along the border with South Africa. (Illustration credit: Simon Lamb)

yourself. In the rest of this chapter—and I will keep doing this in later ones too—I will describe some of the key discoveries geologists have made about geological time and the workings of Earth, discoveries that formed the springboard, as it were, which ultimately launched my own research. All this involves thinking a great deal about rocks. If this at first sight seems unappealing, then let me reassure you that there is light at the end of the tunnel: an understanding of rocks opens up a world that may surprise and fascinate you. But to get there one must first dig the tunnel, and this involves coming to grips with the rocks themselves.

Let me begin by asking what is probably the most basic geological question of all: what precisely are rocks? The answer is simply that they are the solid stuff on which our world rests—the bedrock I have already referred to. For this reason alone, they must be fundamental to our everyday lives, because we live right on top of them. The underlying skeleton, as it were, of every landscape on Earth is made of rocks. If you look closely at these rocks, you will see that they consist of crystals of different minerals, in varying proportions. Some of the ones that might be familiar are calcite, quartz, feldspar, mica, garnet, and olivine, but there are many more. Rocks are clearly not all the same. In fact, there is a huge variety in the continents, and many different ways in which they have formed.

Geologists see Earth as rather like some sort of giant spherical stone fruit, with a thin skin that conceals what lies inside. The skin of this earthly fruit is called the crust, and the underlying interior, made of different rocks than the crust, is called the mantle. The term "mantle" comes from the idea that this part of Earth is a covering for the even deeper core that lies at the heart of the planet, like the juicy flesh around the stone of a fruit. And "crust" is the remnant of the original concept of a solid outer layer resting on a molten interior—a bit like slag floating on molten iron—although today it is recognized that most of Earth's interior is solid. In any case, the crust is not the same everywhere. In the continents, where it is called the continental crust, it is much thicker than it is beneath the oceans; it stands higher on Earth's surface and so forms dry land.

Imagine puncturing the skin of Africa (or any other continent for that matter) by drilling a vertical hole right down until you reach the underlying mantle. This is a good way to get a sample of the rocks in the continental crust. In fact, you would be able to drill down for a few tens of kilometers and still be in this crust. It is how these rocks came into existence that makes them interesting. Some—perhaps most of them—were once molten, forming magma. The magma spilled out during volcanic eruptions, solidifying as layers of lava or ash to form volcanic rock; what was left behind cooled deep within the crust, typically ending up as granite. Geologists call them all igneous rocks, from the Latin word for fire; they were "forged in fire." But the bedrock is not stable at Earth's surface. It is broken up into fragments by the forces of erosion and carried away by rivers, wind, or glaciers.

Eventually, the fragments are left stranded in layers. Living organisms play their part too, leaving behind their calcareous shells and other hard parts. Slowly, layer by layer, the detritus accumulates, gradually compressed to form a new bedrock of sandstone, siltstone, mudstone and limestone. These are sedimentary rocks (from the Latin *sedere*, to settle or sit).

At the elevated temperatures and pressures deep in the crust, the sedimentary layers will harden further, as new minerals crystallize. This sort of "cooking" process is referred to by geologists as metamorphism—the minerals in the rock are metamorphosed from one form to another, and the delicate preservation of fossils is obliterated. If we continue drilling down, pushing technology far beyond the existing limits, then we will finally burst through the bottom of the crust and enter the realm of Earth's mantle. Here, we will encounter a type of rock rarely seen on Earth's surface, full of the green mineral olivine. It is a moot point whether mantle rocks are igneous or the result of metamorphism, or both. Their origin is complex, being in part the original cosmic dust that coalesced and melted to form Earth during the creation of the solar system, but this has also locally undergone subsequent re-melting and metamorphism at the extreme temperatures and pressures inside our planet. Pieces of this rock somehow find their way to the surface, where they are often quarried for decorative building stone. And to geologists they are also extremely valuable, providing a unique window into the planet's deep interior.

THE ABYSS OF TIME

It must have taken time for the bedrock, made up of the various types of rock I have just described, to form. But how much time? Until the early 1800s, it was widely believed that the planet's history had the same length as human history, as written down in the Bible. In 1650, the Anglican Archbishop Ussher had published his analysis of the Old Testament genealogies, adding them up since the time of Adam to arrive at a date for the creation of the world on the evening of October 22, 4004 BCE: this made our planet about 6,000 years old. However, it was clear to engineers and scientists who closely examined rocks that this did not leave nearly enough time to create what they could see. The Scottish geologist James Hutton had already reached this conclusion in 1788. He could see clear evidence in the bedrock near his home in Edinburgh for multiple episodes of profound upheavals of Earth's surface. He recognized strata that had originally been deposited on the seabed, but these must have subsequently been pushed up and tilted, because the layers were now nearly vertical. There had also been time enough for this bedrock to be worn smooth by the forces of erosion, and for new layers of sandstone to be deposited on top and be tilted too. Hutton argued for spans of time that, in effect, had no limit, or as he put it, where there was "no vestige of a beginning—no prospect of an end." One of Hutton's students famously referred to this as "the abyss of time." This diametrically opposing view of the age of our planet, compared to that deduced from a literal reading of the Bible, clearly needed some sort of compromise.

It was not, however, until the British geologist Charles Lyell published his *Principles of Geology* in the early 1830s that the vast vistas of geological time—or deep time, as it is sometimes referred to—were finally revealed to most geologists on a more rigorous basis. In this book, Lyell recognized sequences of sedimentary rocks which contained the fossilized remains of organisms that no longer existed, and he used the occurrence of these fossils to define periods in Earth's history. He also summarized many natural and ongoing processes that are slowly changing the landscape, such as submergence or emergence of coastlines, or shifts in the courses of rivers, or the silting up of river mouths, or the burial of the landscape by volcanic eruptions. To account for the large changes in Earth's surface that these must have caused over time, creating mountain ranges, volcanoes, or oceans, he deduced that Earth was at least millions of years old, and in later editions of his book increased this to hundreds of millions of years. Even this turned out to be much too young, and although not Hutton's infinity, we now know that we are dealing with simply gargantuan time spans: up to billions—thousands of millions—of years. This is because there is a clock that can be used to date rocks.

That clock is radioactivity, a phenomenon discovered in 1896 by the French physicist Henri Becquerel who had observed that the proximity of certain uranium minerals could darken photographic plates. It is a phenomenon that can arouse strong emotions among those who consider it at best harmful, and more likely deadly. But radioactivity is a natural process that occurs in almost all rocks at very low levels, and living organisms have been coping with this process ever since life first emerged on Earth. Just like the sand that runs through an hourglass, the radioactive atoms in the rock dwindle away, emitting radiation as they are transformed into new daughter elements. This decay happens at a predictable rate, which can be measured in the laboratory. By counting the beats, recorded in the accumulation of the daughter elements, geochronologists (those who date geological events) can work out the age of a wide range of rocks—I will talk more about how this is done in the next chapter. It usually means dating when magma either solidified deep in Earth's crust or erupted onto its surface. In many cases, the daughter elements have been slowly accumulating over hundreds of millions to billions of years.

THE HEART OF AFRICA

Dating the rocks in Africa has yielded an extraordinary result. A good way to appreciate this is to imagine taking on the job of coloring a blank map of Africa. You have an artist's palette containing blobs of oil paint arranged around its edge, with colors such as yellow, brown, and blue. Now you start painting. First, you use the blue, and paint a few widely spaced patches covering parts of southern, central and western Africa. Around these blue patches, you paint strips of brown. Finally, you start filling in the remaining gaps with yellow. You have just created an image of Africa that reveals an important geological secret. There are cores or kernels of crust made up of very ancient rocks (the blue in the painting), and these are placed, almost like jewels, in a setting of younger crust (the brown). All this is itself enveloped in even younger rocks (the yellow). So, a journey into the heart of Africa is truly a journey into the deep mists of time.

Let us focus our attention on Eswatini (literally "land of the Swazis" in Siswati), where my project was located, appearing on an atlas as a roughly circular region about 150 kilometers in diameter at the top right-hand corner of South Africa, wedged against Mozambique. This small landlocked kingdom was ruled for sixty years, up until 1982, by King Sobhuza II, widely thought during his reign to be one of the most astute men in Africa. He steered the country, then known as Swaziland, through the shoal waters of Black Africa to the north and apartheid Africa to the south. Sobhuza II died during my second field season, and I still have the copy of the *Times of Swaziland* announcing his death. His successor, the present king, was still at school, and in his subsequent long reign he has brought many changes. In 2018, he officially renamed the country Eswatini, and so I will use Eswatini and Swaziland interchangeably in this book.

Eswatini is a country on two topographic levels, forming a step up to southern Africa's high interior, known as the Highveld. Most of the eastern side of the country is only a few hundred meters above sea level, called the Lowveld. This is a hot, humid region, rife with malaria, with wide plains where wild game roam, although sugar cane fields are now encroaching on this land. The western side of the kingdom is much higher and more rugged, rising over 1,000 meters (3,250 feet) above sea level on the edge of the Highveld. The Highveld is out of the malarial zone with a climate that is less humid and cooler than the Lowveld. And importantly, given the nature of my project, the underlying bedrock is some of the oldest bits of those ancient cores or kernels of Africa. The crust here has shown extraordinary geological resilience, making up part of what is known as the Kaapvaal Craton (craton comes from the Ancient Greek word *kratos*, meaning strength). The craton extends westwards beyond Eswatini, underlying much of northeastern South Africa and eastern Botswana.

The Kaapvaal Craton is one of two geologically very special parts of the world (the other is the Pilbara Craton in Western Australia) because there are sedimentary and volcanic rocks here that are the most well-preserved and extensive relics of Earth's very early surface. These have ages reaching 3.5 billion (3,500 million) years old, or even slightly more. In the next chapter, I will tell you how geologists discovered their great age. To put them in perspective, the solar system is just under 4.6 billion years old. This age is based on the oldest meteorites ever discovered on Earth, which are thought to be remnants of the original material that clumped together to form the rocky planets. In other words, the ancient rocks in the Kaapvaal Craton came into existence less than a quarter of the way through the history of our planet.

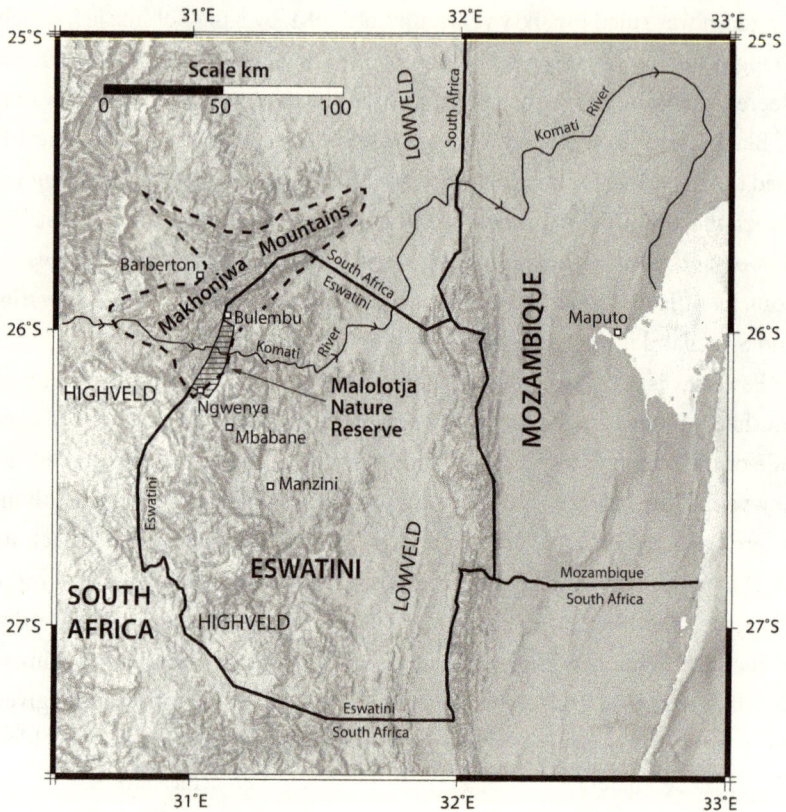

FIGURE 1.2 Eswatini straddles the edge of southern Africa's Highveld, with the western region rising over 1,000 meters (about 3,200 feet) above the Lowveld to the east. My field area was in the Malolotja Nature Reserve, either side of the Komati River and on the southeastern edge of the Makhonjwa Mountains. These mountains are remnants of the Highveld, deeply dissected by rivers flowing to the sea, such as the Komati. (Illustration credit: Simon Lamb)

All this makes these rocks a beacon to geologists, drawing them ever closer to tangible remnants of our young planet. The best place to look is in the Makhonjwa Mountains, which locally rise over 1,800 meters (5,800 feet) above sea level, straddling the international border between Eswatini and South Africa. And so it was in the Makhonjwa Mountains, where they lie in the northwest corner of Eswatini, that my quest to understand the early years of our planet began. It turned out that much of my research was in a newly created conservation area called the Malolotja Nature Reserve.

A TIME MACHINE

The task of looking across geological time might seem a daunting undertaking, but it is something geologists are doing all the time. If you walk through the geological galleries of a major natural history museum—say, the Smithsonian in Washington or Natural History Museum in London—then you might well encounter a series of displays of what geologists have come up with, providing an astonishingly rich vision for parts of our planet's past. As a brief example, here are some snapshots of the last few hundred million years:

4,000 BCE: Not much has changed except that, even if you look closely, there are no obvious signs of human activity. Large tracts of land are dense forest, with no roads, no cities in most places, no sprawling patchwork of giant fields, and on the dark horizon, no lights.

18,000 BCE: Huge expanses of white cover the northern hemisphere, smothering Canada, and much of northern Europe. In the great ranges of the European Alps and Rockies, only small, isolated rocky peaks poke above the white. This is the height of an ice age, when ice sheets spread southward from the North Pole. Sea levels are over one hundred meters lower than today, and there are wide plains along the fringes of continents, with many islands now connected to the mainland.

5.6 million BCE: The Mediterranean is somehow missing—instead of the familiar blue, you see a salt pan, left behind when the Mediterranean dried up.

66 million BCE and beyond: The continents are no longer in their familiar positions: the Atlantic is too narrow, and South America and Africa have peculiar shapes. There is no ice anywhere—Antarctica is green, cloaked in dense forest. And where are the high and rugged ranges of Tibet, Himalayas, and the Andes? Most of what you can see are green plains or sea. Dinosaurs and giant reptiles roam the surface. Going further back in time, the outlines of the landmasses seem to be unstable, shrinking and expanding like some pulsing organism. The landmasses themselves are also constantly moving. Sometimes they coalesce into a single continent, and the rest of the world is ocean—the date is 250 million years BCE—and sometimes they are dispersed all over the planet. The vegetation on land changes drastically too, and eventually much of the land is barren.

I suspect that most museum visitors would take these snapshots of our planet in the past for granted and think no more about them. In fact, they have just viewed the results of a vast amount of scientific research, especially because geologists have access to vastly more detailed information than these descriptions would suggest. So, what made this possible? The answer, of course, is rocks. One of the great achievements of geologists, alongside the recognition of geological time, is the discovery that rocks are not just the bedrock of our world—varied though they are—but also a time machine, making it possible to look into our deep past and see ancient landscapes, vanished oceans, long-lost mountain ranges, extinct life forms, climate change, natural disasters, and the interior workings of the planet. The record is extraordinarily wide-ranging and detailed, and almost immune to the effects of time. It is, however, in a strange code that must be deciphered before much sense can be made of it. Cracking this code and understanding the story of our planet is the science of geology.

READING THE ROCKS

The creation of each sedimentary or volcanic layer in the bedrock would probably make news headlines today, because it is likely to be the result of a natural disaster. It could be what was left after the flooding of a river, or a period of violent windstorms, or a retreating glacier, or the eruption of a volcano, or landslides triggered by an earthquake. More fundamentally, the layers mark incremental changes in the landscape.

Over time, the layers may become buried as more accumulate on top. Eventually, they are exposed again when, after being pushed up and tilted by forces within Earth, they are subjected to erosion. In this way, they are revealed like pages in a book that geologists have learned to read. It is hard to know where to start because the record is so rich. By looking at the shapes and sizes of the individual sand grains, and the way they build up the rock, geologists can single out the natural disaster that was responsible. As I will explain in later chapters, they will also be able to visualize this disaster in great detail. If it was a river in flood, then they will know which way it was flowing, and possibly how fast; for a sandstorm, the prevailing wind direction. An ash cloud from an erupting volcano will leave behind clues to how it engulfed the landscape. And if there are fossils in any of these layers, then their particular adaptations might also reveal the ancient environment in which they lived.

Although not directly observable in the field, rocks contain a wealth of information in their mineral and chemical composition. As a result, most fieldwork involves collecting samples for further study back in the laboratory. A day in the field is usually a day lugging around an increasingly heavy backpack, as it steadily fills up with rock samples! Geologists have developed clever laboratory techniques for analyzing these samples, winkling out remarkably precise details about the geological past. It might be possible to date some of them using the radioactive clock. For rocks that were once deep inside Earth, one can deduce how they got there, and how they ended up back at the surface, by probing the minerals they contain. And if you are interested in conditions at Earth's surface, then one can analyze the composition of sedimentary rocks that have precipitated from surface water. This composition can be linked to such things as the temperature of the ocean, the climate and concentration of atmospheric gases, and even altitude.

The record in the rock layers only makes true sense if read in the correct order. However, our planet does not normally yield up complete "books" or even "chapters," but instead odd pages. Geologists routinely attempt to group as many pages as possible into hierarchical schemes: systems, groups, formations, and the like. However, working out their geological ages has proved much more difficult, because sedimentary rocks are difficult to date using the radioactive clock. Here, the evolution of life has come to the aid of geologists. For the past 540 million years or so, living organisms have been sufficiently diverse that it has been possible to use their fossilized remains to bring some sort of order. Through painstaking work examining the fossils preserved in the rock layers and relying on the fundamental principle that the evolution of life never precisely repeats itself, geologists have worked out how to assign the right chapters, and sometimes even page numbers. This way, they have carved up time, in much the same way that archaeologists or historians have done with human history—but obviously, over far longer periods—giving names to past chunks of geological history that can be directly linked to the fossils that are found in the rocks.

Consider the time of the dinosaurs. From the point of view of a geologist, they lived in what was once thought to be the "time of fossils," called the Phanerozoic Eon (*phaneros* is Ancient Greek for visible, and *zoic* for life), more precisely in the Mesozoic Era ("middle age" of life), and specifically during the Triassic, Jurassic and Cretaceous periods (named after localities or distinctive features of these rocks). For particular types of dinosaurs, the intervals of time when they emerged or became extinct can be narrowed down to smaller units called "epochs" and "ages," with the finest scale resolution of a few million years.

Unfortunately, as we go back further in time, we no longer have enough fossils to subdivide time this way, and it becomes more and more difficult to make sense of the rock layers and put them together into a coherent history. Therefore, a major challenge facing any geologist studying such ancient rocks is fitting the layers all back together in the order in which they were laid down so that the rock record can be deciphered. This, as you will see, is what this book is all about.

- § -

GIANT PLATES

Finally, I get to plate tectonics. I am always slightly uneasy about using this odd-sounding phrase, because I have this feeling that it reinforces in many people's minds the idea that geology is an obscure subject (and perhaps also dull)—but this could not be further from the truth. Plate tectonics is one of the defining features of our planet today, and an understanding of how it works was a critical part of interpreting the very ancient rocks in my study area in the Makhonjwa Mountains of Eswatini and South Africa. So, I should explain what plate tectonics is, and I think the best way to do this is by telling the remarkable story of how it was discovered, back in the 1950s and '60s.

As often happens in science, serendipity played an important role, in this case, through growing military demands during the Cold War between the former Soviet Union and America and its allies. The oceans were now the hunting grounds of nuclear submarines, and it became imperative to produce detailed maps of the seafloor—up until then the surface of the moon was known better than this undersea world in which submarines played cat-and-mouse games. Scientists embraced these new opportunities, exploring the seafloor with echo sounders and other geophysical techniques. Once they had, in effect, whisked away the covering of water to reveal what lay underneath, they found a strange new landscape. There were continuous mountain chains—the mid-ocean ridges—running down the middle of the oceans, with active volcanoes along their crests. Numerous fault lines—now known as transform faults—appeared to regularly offset the mid-ocean ridges by tens to hundreds of kilometers, creating a sort of zigzag pattern in the seafloor. Ocean trenches, about 50 kilometers wide, formed deep and arcuate grooves almost right round the Pacific Ocean and along the edge of the Indian Ocean, reaching nearly 11 kilometers (6.8 miles) below sea level. This peculiar image of the seafloor set the scene for the idea of plate tectonics.

One of the key breakthroughs, however, came about through the need to monitor the testing of atomic bombs during the Cold War nuclear arms race. In 1963, after many years of negotiation, the Limited Nuclear Test Ban treaty was signed in an attempt to de-escalate political tensions. For treaty obligations to be verifiable, scientists had to be able to detect a nuclear explosion, which involved distinguishing it from a natural earthquake. This prompted geophysicists to develop new techniques to measure seismic vibrations, and importantly, work out the location of their source and the movements in Earth that caused them. Geophysicists were exploiting the fact that the shaking during an earthquake is quite different from that caused by a nuclear explosion: the former is the result of sudden movement along a fracture or fault in a rock, whereas the latter is an outwardly directed blast in all directions, setting off a different pattern of vibrations.

A worldwide network of seismometers was set up to look for signs of atomic testing—in effect, listening posts for the smallest vibrations in Earth. This turned out to be like searching for a needle in a haystack because the network was soon swamped by hundreds of thousands of natural earthquakes. And it became obvious that the locations of these earthquakes were not random. There were large parts of the world that were virtually earthquake-free, ringed by narrow zones of intense earthquake activity. Because earthquakes are a sign that the rocks are breaking up along fault lines, earthquake-free regions are likely to be where the rocks are remaining intact and rigid. These regions soon became known as the plates, and the intense zones of earthquake activity marked out their edges.

In 1967, in a landmark scientific paper, Dan McKenzie and Robert Parker demonstrated a remarkable simplicity to the large earthquakes around the margin of the Pacific, in a region extending from southern California to Alaska, and on to Kamchatka and Japan—a distance of over 10,000 kilometers (6,250 miles). They calculated the fault movements that had caused earthquakes, and in almost every case, these were in directions that that lay on the arcs of circles, rather like lines of latitude. But these lines did not ring the geographic poles; instead, they were centered on a pole elsewhere on the surface of Earth. McKenzie and Parker realized that this could be explained if the rocks were breaking along the junction between two rigid portions of the globe—now recognized as the Pacific and North American tectonic plates—as they twisted relative to each other, exactly as would be anticipated for moving plates on a spherical earth.

Today, thanks to the major advances in the detection and analysis of earthquakes, McKenzie and Parker's idea has been confirmed many times over. The world has now been divided into fifteen major plates and their motions can be tracked directly from space, on a near-daily basis, using an advanced version of the same GPS (global positioning system) technology that we have in our mobile phones. The speed of a plate is incredibly small by the standards of everyday life, only a few tens of millimeters (a few inches) every year—a typical garden snail moves about a million times faster than this. But over the colossal span of geological time—many hundreds of millions of years—this can add up to thousands of kilometers.

MAGNETIC FLIP-FLOPS

There are more consequences to plate tectonics than just earthquakes at the edges of the moving plates. The crest of the mid-ocean ridge is where plates drift apart. It is also a place of intense volcanic activity, erupting a lava known as basalt. The significance of this first emerged in the late 1950s after geophysicists had begun measuring in detail the strength of Earth's magnetic field in the oceans.

We all know that Earth's magnetic field gives us north and south magnetic poles, which we rely on when we use a compass to find north. It is less well known that the strength of this magnetism decreases overall from the poles to the equator. In fact, magnetic measurements show in detail that it is more complicated than this. Magnetic surveys on land, in the continents, have long revealed a very irregular pattern of highs and lows, defining blob-like patches where the field is slightly stronger or weaker than normal—a consequence of the variable magnetism in the wide range of underlying rocks. But to everyone's surprise, subsequent measurements taken in the oceans produced an astonishingly regular pattern of magnetic highs and lows. When plotted in black and white on a map, with black for a magnetic "high" and white for a magnetic "low," they created zebra-like stripes of so-called magnetic anomalies running parallel to the mid-ocean ridge. Even more remarkably, the pattern of anomalies was virtually identical on either side of the mid-ocean ridge crest.

In 1963, in a short paper published in *Nature*, Fred Vine and Drummond Matthews at Cambridge University proposed an explanation. They had a hunch that the magnetic anomalies reflected variations in the magnetism of the volcanic rocks on the seafloor. They knew the basic principles of rock magnetism: when a lava solidifies, it becomes slightly magnetic and behaves a bit like a bar magnet or compass pointing to Earth's magnetic north pole. They knew something else: the magnetic north pole was extremely unstable and had even, on occasion, completely flipped or reversed, so that a compass needle would point south rather than north. The last time that this had happened was about three-quarters of a million years ago, but there had been many other flips before this.

Vine and Matthews's great insight was to put all these ideas together. If an eruption of lava on the seafloor happened to cool during a period when

Earth's magnetic field had undergone one of its flips or reversals during which magnetic north became south, the rock would also acquire a magnetism in the opposite direction. They calculated how reversals like this in the magnetism of rocks on the seafloor would affect the magnetic field at the sea surface, and there was an almost perfect match with the ocean magnetic surveys. The seafloor seemed to be behaving like a tape recorder, picking up the flips in Earth's magnetic field as a series of stripes. Vine and Matthews reasoned that this could only happen if the crust beneath the sea floor was being continually created by volcanic eruptions at the crest of the mid-ocean ridge and then imprinted with the direction of the magnetic pole before cooling and steadily moving away on a giant conveyor belt of crust. This conveyor belt was part of a deeper flow in the mantle, giving rise to what has become known as seafloor spreading. It was subsequently realized that the cooling rocks would also contract slightly and subside as they moved away, making the ocean progressively deeper. This is what actually creates the topography of the mid-ocean ridge, and its crest is literally the birthplace of the seafloor.

MEGATHRUSTS AND BARNACLES

Seafloor spreading provides a mechanism for plates to drift apart almost without limit. On its own, however, it does not necessarily lead to plate tectonics, and in the early days of its discovery, there were geologists who thought it was proof that Earth was expanding, increasing the planet's surface area. It was not until the full significance of the deep ocean trenches was understood that seafloor spreading could be fitted into the theory of plate tectonics. Again, chance played a role: not long after the idea of seafloor spreading had been proposed, and with the newly established worldwide seismic network in place, a magnitude 9.2 earthquake struck Alaska, near Anchorage, on Good Friday 1964. The disaster zone was on the northeastern margin of the Pacific, where an ocean trench lies about 200 kilometers offshore. A young geologist named George Plafker was sent by the United States Geological Survey to investigate what had happened. He found that a profound change to the landscape had occurred during the few minutes of the earthquake. A vast tract of land, extending for over 700 kilometers along the Pacific coast of Alaska, had been uplifted out of the sea up to 12 meters (40 feet), exposing parts of the seabed. Another large area farther inland had subsided up to 2 meters (6 feet), and the subsequent inundation of the sea flooded extensive coastal regions.

George realized that the sheer size of the affected region implied the rupture of a gigantic fault. And to explain the land-level changes, the fault must underlie the whole region, forming a gentle ramp that reached the surface offshore in the deep ocean trench. He called this fault a megathrust. During the earthquake, sudden movement along the megathrust had resulted in the seafloor at the bottom of the trench being shoved beneath the continental margin of Alaska, jacking up the coastal regions and tilting them landward. George's idea linked ocean trenches with geological activity on land for the first time. And the earthquake was clearly not a one-off event. There were signs that something like this had occurred many times before, creating in the region of uplift a series of steps in the landscape like a giant staircase ascending from the coast; each step was formed during an earthquake as the bedrock was raised out of the sea.

In 1996, I had the opportunity to spend several weeks with George, revisiting many of the places around Prince William Sound where he had made his key observations in the summers of 1964 and '65. He had warned me that after thirty-two years we might not find much, because what had been so fresh in the immediate aftermath was probably covered up by dense vegetation or eroded away. So you can imagine how excited we were to discover that everywhere we looked, the evidence for the earthquake was still visible and preserved in exquisite detail, also giving us an opportunity to see if land levels had changed in the following decades in the buildup to the inevitable next earthquake. Along the rocky shoreline, in the region of uplift, there were stranded barnacles still clinging to the rocks, but several meters above the mean high tide mark where they would have originally grown. I remember George turning to me as he picked at a barnacle with his finger, remarking that some of these were still alive in the seconds before the earthquake, only to be yanked out of the sea to their present position. He had used the same barnacles to estimate uplift during the earthquake. In the region of subsidence, we photographed forests of dead pine trees, killed when the sea invaded the low-lying coastal plains in the immediate aftermath. Elsewhere in Prince William Sound, high up on hillsides, there was even wreckage lodged amongst the branches of pine trees, left behind by one of the giant water waves that had swamped the coast. I saw poignant reminders of the human tragedy, such as a child's shoe, an enamel bowl, or a tangle of fishing line.

George Plafker's discovery of the megathrust beneath Alaska led directly to the idea of subduction. Subduction occurs when a tectonic plate slips into the underlying mantle along a megathrust, and it happens in a subduction zone. Subduction zones extend around most of the Pacific margin, wherever there is an ocean trench offshore, and they are regularly shaken by great earthquakes due to rupturing of the underlying megathrusts. Repeated displacements on a megathrust over millions of years allow the plate beneath the ocean to slide hundreds of kilometers into the underlying mantle, pulling down the sea floor to create the deep ocean trench—a trail of earthquakes tracks the sinking plate as it descends to depths of nearly 700 kilometers. But we live on a nearly spherical planet, and so wherever a plate is pulled down like this, the resultant depression follows the arc of a circle—try making a dent in a table tennis ball—and this explains why the ocean trenches are distinctive arcuate grooves on Earth's surface. This way, the sea floor created at the mid-ocean ridge is returned to Earth's interior, eventually closing up an ocean. But the sinking plate, as it heats up again in the mantle, triggers new volcanic activity. Volcanoes along subduction zones stand like sentinels parallel to the deep ocean trenches, erupting a distinctive lava called andesite after the volcanoes in the Andes where there is also a subduction zone. These are the Ring of Fire, and also the volcanic chains through Southeast Asia, forming a series of arcuate volcanic necklaces that encircle our world.

BUILDING THE WORLD

Finally, with most of the pieces of the puzzle in place, we can start to see the full picture of plate tectonics. The surface of Earth turns out to be far from fixed, but instead, it is made up of a mosaic of restless plates. The plates go their separate ways, like gigantic bumper cars in some fantastical fairground attraction: sometimes coming together as one plate sinks beneath another in a subduction zone, sometimes moving apart and opening up a new ocean,

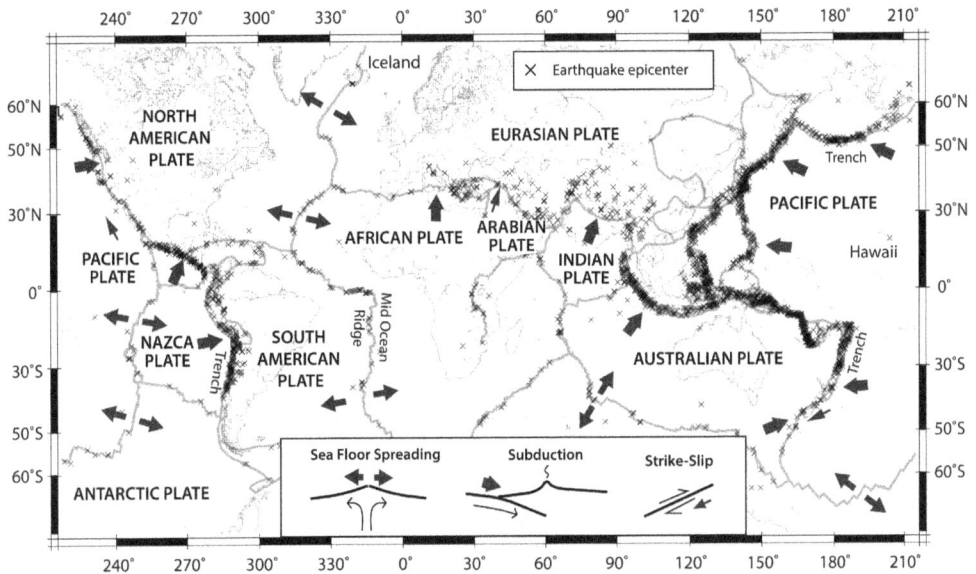

FIGURE 1.3 Earth's surface is a mosaic of tectonic plates that move apart at sea-floor spreading centers, slide past each other along strike-slip faults, or converge in subduction zones or regions of continental collision. Plate boundaries are regions of intense earthquake activity. (Illustration credit: Simon Lamb)

and sometimes just sliding past each other. The continents are carried along with the plates, wandering over the globe, in the process rearranging and reshaping our world. This is the significance of the "tectonics" bit, taken from the Ancient Greek word *tekton*, meaning builder or mason, because it is all about building the outer part of our planet. It turns out that this only applies to the top hundred or so kilometers of the planet. Deeper down, below the plates, the rocks are so hot that they can actually flow over geological time like a fluid. This ability to flow has many profound consequences for the deep workings of Earth, which I will come to in later chapters.

A natural question that soon arose was: how long had plate tectonics been going on? Given its role in the workings of our planet today, this is no idle question; it is central to our understanding of the evolution of our planet. Was it something that emerged relatively late in Earth's history, in a planet that was previously more akin to Venus or Mars, where no clear signs of plate tectonics have been found? This would require a sort of terrestrial midlife crisis, resulting in a fundamental change in behavior. Or was this all sorted out in Earth's very early childhood? These questions go to the heart of a deeper philosophical approach to our planet: the principle of uniformitarianism, first put forward by Charles Lyell in the 1830s in his seminal *Principles of Geology*. In fact, Lyell gave the first edition of his book the subtitle "An attempt to explain the former changes of the Earth's surface by reference to causes now in operation." This is sometimes summed up in the catchphrase "the present is the key to the past." In other words, the forces or mechanisms within Earth that drive geological activity today are the same as those in the past, and they have therefore remained "uniform"—that is, essentially unchanged—over geological time.

When Lyell first argued this concept, it was considered a new vision of our planet that radically departed from the widespread belief that the main features of the planet were created during the catastrophic event of the biblical flood, at the time of Noah. However, Lyell's focus was on what geologists would regard today as relatively late stages in our planet's history, during the past half billion years or so. The discovery that our planet is many billions of years old has required a new look at the idea of uniformitarianism, especially as we now know that Venus and Mars must have evolved in quite different ways. This is where my project in southern Africa comes in, because my thesis supervisor, Alan Smith at Cambridge University, had been a pioneer of using the new ideas of plate tectonics to make sense of rocks. He had realized that the rocks in the Makhonjwa Mountains might reveal whether the concept of uniformitarianism could be extended back to early Earth, and in particular whether plate tectonic activity was occurring at that time.

A NEW PROJECT

Back in the 1970s, Alan Smith was running a project studying much younger rocks that were thought to be fragments of oceanic crust. These were collectively known as ophiolites—from the Ancient Greek *ophis*, or snake—because they often contained a scaly and slippery mantle rock known as serpentinite (also meaning snake-like, but this time in Latin—nineteenth- and twentieth-century geologists liked to show off their knowledge of the classics). Given where these ophiolites were found in the Alpine-Himalayan mountain chain, Alan realized that plate tectonics could explain them as relics of the floor of the Tethys Ocean (Tethys being the wife of the Greek god Oceanus). This ocean had once lain between Europe and Africa, extending eastward between Asia and India. About 90 million years ago, Africa and India were far to the south of their present positions; they had subsequently drifted northwards as the floor of the Tethys Ocean slid back into the mantle in a subduction zone. The rise of the Alpine-Himalayan mountain chain was the final act in the closing-up of this ocean.

Alan had several students studying ophiolites in the mountains of northern Greece. They were trying to work out how the oceanic crust, which had once lain beneath the deep seafloor, had ended up on dry land—and what this revealed about the dying moments of the Tethys Ocean. One of these students went on to work for the former Swaziland Geological Survey as part of an aid program with the British Geological Survey, helping the small kingdom explore its own mineral wealth. As part of this, he had visited the Makhonjwa Mountains where they lie in northwestern Eswatini and immediately realized that there were similarities between the rocks here and the ophiolites he had been studying in Greece. By now it was known that these rocks were around 3.5 billion years old. So, could these be some of the oldest fragments of seafloor known on Earth, created through sea floor spreading and then pushed up on land during the closing of an ancient ocean in a subduction zone, perhaps when two continents finally collided? In other words, was this some of the oldest evidence for plate tectonics? It had even been speculated that such early plate tectonics had played a role in getting life started. Amongst geologists, these were pretty exciting questions.

Alan Smith agreed to supervise a student who would study the ancient ophiolite-like rocks in Swaziland. This was how my doctoral project was born. And as you will soon find out, the project quickly shifted away from just ophiolites to become a much bigger geological detective story.

GREEN LIKE THE SKY

Now that you have been introduced to many of the fundamental ideas in geology, you should be well versed in the rationale for my research project. However, I think one can also take a much broader view of its relevance that goes beyond just geology. Let me therefore end this chapter by considering a curious phrase in the Siswati language of Eswatini that to my mind is an elegant way of thinking about our planet. How I came across this bit of Siswati is a story in itself—a story that certainly illustrates the practical problems of carrying out fieldwork in remote places. I have described in more detail what happened in the anecdote at the end of the chapter. But for the moment, all you need to know is that on my first visit to Swaziland I narrowly missed being killed when my vehicle, privately hired from a resident British couple, rolled down a steep hillside.

The following day I returned to the scene of the accident with Helen, one of the vehicle owners, to see if we could talk to a boy who had witnessed the accident. Speaking haltingly in Siswati, she asked the local farmer if he knew the whereabouts of the boy. The old man seemed to follow what she was saying up to a certain point and then became puzzled and perplexed. The stumbling block was the blue color of the boy's T-shirt. When I understood this myself, I looked around for something of the same color, including my blue jeans (unfortunately, the wrong shade of blue). The farmer watched us closely as we discussed in English suitable blue objects. The word blue must have come up again and again in our conversation, because finally he used the English word "blue" himself when saying in Siswati: "The boy with the 'blue' T-shirt? You want the boy with the 'blue' T-shirt?" Helen laughed with relief. "Yes, yes, that boy." But then he shook his head, indicating he could not help us; the whole conversation had come to nothing.

Later I asked Helen why she had had such difficulty communicating with the farmer. 'Well," she said, slightly puzzled, "I used the Siswati word for blue that I had been taught during my language course. My teacher told me that there was no specific word, so when referring to this color, Swazis say a phrase which would be rendered into English as something like 'green like the sky.'" But it seems not everybody is familiar with this elaborate phrase, or perhaps it varies in its details amongst different groups of speakers, or perhaps Helen's pronunciation was the problem. In any event, my encounter with this strange form of words had a lasting effect, and I have thought about it ever since.

"Green like the sky" was Helen's translation of the Siswati words "luhlata lwesibhakabhaka" (*uluhlata* means green, also green grass; *isibhakabhaka* means sky). As far as I can tell, "green from the sky," "green of the sky," or just "sky green" are all possible translations too—in fact, any association between green (and green grass) and sky. A friend of mine who knew Swaziland well once told me that Siswati names often reflect ideas that can be difficult to translate concisely into English. I like to think that there is more meaning here than just the fact that the daytime sky is usually blue. Interpreting this from the perspective of the traditional Swazi way of life, dependent on the hilly grasslands to feed the wealth-giving herds of cattle, the sky can also *make* green—thereby directly connecting these two colors—by providing water in the form of rain that fills the rivers and stimulates new growth in the landscape. I have witnessed many such greening events during my time in the Highveld of Swaziland, brought on by the rains after the spring burn-offs of the long brown grass. And when the rain has gone, the blue sky returns above the newly greened hills. This way, blue and green are juxtaposed with clear cause and effect. Space exploration has given us a fresh perspective on these distinctive colors of our planet, allowing us to look down through the sky. We still see green vegetation, but the sky is now transparent except for the swirls of moisture-laden white clouds and a thin bright blue horizon. Most of the blue in this image, covering nearly two-thirds of Earth's surface, is the oceans. As a scientist, I find many surprising interconnections between these observations. A bit of lateral thinking gives you even more.

Life exists today because Earth has an atmosphere that cocoons us in space, allowing liquid water to exist on the planet's surface. This atmosphere is the reason the sky looks blue: it contains gases of oxygen and nitrogen that scatter sunlight strongly in the short wavelengths, singling out the blue end of the color spectrum. All the important greenhouse gases that keep the atmosphere warm and make life possible—water vapor, carbon dioxide, methane, nitrous oxide, and oxygen as ozone—are exchanged with the atmosphere when plants capture sunlight and undergo photosynthesis and respiration, and when they die. The most widespread organisms that do this are cyanobacteria, also known as blue-green algae. The oceans are blue because they absorb strongly in the red and infrared end of the spectrum, reflecting short-wavelength blue light. And this behavior is the defining characteristic of a greenhouse gas.

It has long struck me that "Green Like the Sky" would be a perfect subtitle for Earth, a form of shorthand for the essential features of our blue planet. Perhaps this could have been added to the plaques that went on the Pioneer space probes, or to the gold-plated records on the Voyager missions in the 1970s, put there in the forlorn hope that they might be found by intelligent alien life. It certainly would help these aliens find Earth today—if they don't wait until we have destroyed much of the blue and green through our current behavior. But what about billions of years ago, when the solar system was very young? What sort of planet would Earth have been back then? Would the epithet "Green like the Sky" still have applied? These questions lie at the heart of my book, and my own search for answers begins with the experiences I had as a geologist in Eswatini, using its rocks to look back into the depths of time. The old gold miners in southern Africa called the rocks that I was looking at 'greenstones,' simply because they often had a greenish color. Over the course of my research career I have returned to these greenstones many times, either in fact or thought.

In the next chapter, I will tell you how geologists discovered and began to read the ancient rocks in southern Africa, and how I got started in the Makhonjwa Mountains. As a sort of coda, I will end each chapter with some short stories about my experiences in southern Africa.

- § -

AN ACCIDENT WAITING TO HAPPEN

I begin deep in a narrow and rocky valley, lost on the Highveld of what was Swaziland in southern Africa, where the following scene unfolds. From afar, it seems as though the group of people are stiff and mute. But much closer up, glimpsed at through the thorny scrub and stumps of hacked-down acacia and protea bushes, their conversation can be clearly overheard, alternating between the hesitant intonation of somebody trying to communicate in a foreign language and the click sounds of a fluent Siswati speaker. They are standing on a bare-earth ledge on the hillside, in front of a circular reed hut. Chickens cluck and strut past with jerky movements, and smoke rises from the open hearth. A tall thin man—a local farmer, with grey hair and a wispy beard—leans on a long stick with a perplexed expression on his face. A teenage boy—the man's son, perhaps—stands awkwardly, looking blankly with his hands locked together behind his back and his body twisted half round. A large woman bends over the fire, with two young children clinging to her skirt.

In front of them, a woman is speaking loudly, gesticulating and pointing. Next to her, a young man with a bright red beard wearing blue jeans, a long-sleeved red check shirt, and a white floppy sun hat says nothing but looks grim and worried. The conversation seems to be repeating itself, like a scratched record. Each time the woman finishes, the old man shakes his head and replies questioningly. At that, the woman launches again into her limited Siswati phrases and the whole cycle begins again. Finally, the man with the red beard and floppy hat bursts out in English: "What is he saying?" "He says he can't understand me," the woman replies.

Who is this young man with the bright red beard? It is me nearly forty years ago. And what am I doing here? For the moment, I will just go back nearly twenty-four hours, to the day before, although you also need to know that I am a geologist. On that day, around eight in the morning, I had made the decision to visit an abandoned mine where unusual layers of the mineral barite can be found in the rocks on the steep side of the valley. Getting there involved following a long and tortuous rough track. I had been told all this by a local government geologist. But whether because it had completely slipped his mind, or because I was not listening attentively enough—I am still not sure which—I failed to appreciate that this track was only suitable for four-wheel drive vehicles, and certainly not for the ancient yellow Volkswagen camper van I was driving. I had hired the van from a British couple—the wife's name was Helen—who planned to drive home to England overland through Africa. The husband was very ill with a high fever, and this was delaying their departure. And the Volkswagen was ill too, close to the end of its life: the steering wobbled, the engine lacked power, and the brakes were soft. Even without these weaknesses, this vehicle should certainly not have been on the road to the barite mine. I realized this shortly after I had left the comparative safety of the main sealed road and turned off onto the old dirt road.

The track continued for several miles. It became steeper and more deeply rutted; the recent rains had turned it into two deep parallel ditches. I found myself trying to steer a course with the wheels perpetually perched on the edges of these ditches. My hands were sweating, slipping on the rim of the steering wheel. Every so often, the undulations in the track would twist the front wheels, wrenching the steering wheel out of my hands. There was no room to turn around, and the edge was becoming ever more precipitous. As I rounded a sharp bend, I saw the track ahead plunge away alarmingly and I knew now that I would never be able to get back on my own. I desperately needed help. And here a new problem reared its ugly head. With the engine off, the hand brake was not strong enough to hold the vehicle—I could only just manage this by pressing hard on the foot brake. Somehow, I had to get out and wedge some stones under the wheel.

At this point I saw a young boy standing beside the track, wearing a bright blue T-shirt and watching me intently. I tried to indicate to him that I needed to chock the wheels. And in my frantic efforts, my foot suddenly slipped off the foot brake and the vehicle started rolling forward, faster and faster. In no time at all I had lost control. I had no choice. I opened the door and jumped out, grazing my leg on a sharp stone. The camper van bounced away from me and down the hill, gathering speed all the time. Not more than a few tens of meters away, the wheels must have hit a rut because it suddenly lurched off the track and tilted over, rolling down the steep hillside, turning over and over. The doors flew open and the contents of the car spilled out, leaving a trail of debris along the van's last journey. It fell about fifty meters or so before crashing into a streambed at the bottom of the valley, coming to rest on its side with two of its wheels spinning uselessly in the bright morning sunshine.

As I looked down at the wreckage, I tried to make sense of what had just happened. I spotted some of my field equipment—my field notebook, maps, and aerial photographs—caught up in bushes and long grass. Retrieving them and turning my back on the disaster, I started the long walk up the track to the main road. The boy had gone, but I hardly noticed because I was so occupied with how I was going to tell the British couple what I had just done to their van. I decided to get it over with it as quickly as possible. To my relief, they were far more concerned about my own safety than what had happened to the van. But Helen did want to visit the scene of the accident and find the boy with the blue T-shirt, who might be a helpful witness for an insurance claim. This was why I found myself back near the barite mine the next day, listening to the strange conversation in Siswati.

2 | Makhonjwa Mountains

First encounters with ancient rocks

On the wall of my living room is a black-and-white photograph of the Makhonjwa Mountains in Eswatini. I took the picture during one of my field seasons, developing the film and making the print myself, and so the image is very personal to me. The mountains appear as jutting craggy masses or ridges with bare grassy sides, dotted with stands of protea trees and thorn bushes. The lower slopes are usually truncated by high cliffs where rivers have cut deeply into the bedrock. This has created thickly vegetated and virtually impassable grooves across the landscape in which the rivers lie hidden from view: the surrounding peaks rise up to 1,000 meters (3,250 feet) above the riverbeds, giving them a scale similar to the Scottish highlands. However, because the river valleys are also carved out of the edge of South Africa's Highveld, one actually looks down on the high peaks when approaching on some routes from the continental interior. Crossing these mountains on foot is an exhausting undertaking; you must scramble up hot stony hillsides and along undulating ridges, avoiding the deep gullies. You have to force your way through long grass and prickly scrub, half blinded by the glare of the sun reflected off the rocks. But it all seems worthwhile in the late afternoons when the hills have a kinder and calmer feel, tinted a warm orange and molded by the shadows. The only sound is the wind in the grass, or the muted roar of water cascading down the gorges, and the occasional screech or bark of a monkey.

Whenever I look up at my photograph, I feel as though I am being transported to a mysterious hidden world, lost in the enormity of Africa and waiting to be explored. It is not hard to imagine that only a century or so ago these mountains were unknown to science; nobody had an inkling that there was anything special about the bedrock here. But once geologists started to explore this part of southern Africa, the mountains rapidly emerged from scientific obscurity to become a place of extraordinary interest to those seeking to uncover the early history of our planet. My own first visit here was just the latest stage in this endeavor. I will begin the story, however, much farther back in time.

FIGURE 2.1 Rugged hills in the northern part of my field area in the Malolotja Nature Reserve, Eswatini, underlain by sedimentary and volcanic rocks dating to 3.2–3.5 billion years ago. The deep Komati River gorge is faintly visible, running across the middle of the image. (Photo credit: Simon Lamb)

EARLY DAYS

About forty thousand years ago, hunter-gatherers crossing the Ngwenya Hills, on the southeastern edge of the Makhonjwa Mountains, noticed a red stain in the soil and scooped out a rust-red powder of iron oxides. Mixing this with water, they became artists, most likely decorating themselves and their rock shelters with strange patterns. They and their Stone Age descendants also left behind signs of their everyday lives: hand axes, cleavers and stone arrowheads can be found in the banks of rivers that traverse the region. But it was not until much later, when the ancestors of the Swazi people arrived from the north and displaced the lions, giraffes, elephants, and antelope, that permanent settlements were established. These people built clusters of round huts made of mud and grass and grazed their herds of cows on the hillsides.

During the nineteenth century, the region saw an influx of missionaries, settlers and speculators of European origin. Dutch nomadic farmers known as Boer Afrikaners, who had long lived in the Cape region much farther south, migrated northwards and carved out new territories that included the Makhonjwa Mountains. They came into conflict with both the indigenous inhabitants and the British Empire. The British sought to establish their own colonies during the great scramble for Africa, in a drive to exploit the continent's rich mineral wealth. They had been lured by the recent discoveries of gold and diamonds. After many clashes, the British defeated the Boers in the Boer War, which ended in 1902. Soon afterwards, the Makhonjwa Mountains were deemed to mark the eastern edge of the Transvaal colony (later becoming the Transvaal Province, known since 1994 as Mpumalanga) and the western edge of the British protectorate of Swaziland, jointly ruled by the British government and the Swazi monarchy until independence in 1968.

During the First World War, a British geologist named Arthur Lewis Hall set up camp in the Makhonjwa Mountains. This is when the history of the region comes into much sharper focus for me. It turns out that I have a connection with Hall: he also studied geology at Cambridge University and, like me, had been awarded the Harkness Scholarship—I was using some of the money from this scholarship to pay for my own fieldwork in these mountains. Hall's stay in the Makhonjwa Mountains is also preserved in photographs, and I have camped in similar places. One can see him standing at the entrance to his tent and squinting into the rays of the setting sun. Nearby, a horse tethered to a tent peg munches amiably, its nose deep in a canvas bag of oats. The tent is a brilliant white conical structure, held up by a central pole and gripping the ground with its radiating guy ropes. And in the surrounding landscape of grass, protea bushes cast long shadows that streak the hills with dark lines.

Hall was the first geologist to survey these mountains. Prospectors had already worked their way up the valleys, panning for gold and sparking a gold rush to the area. Now, the government in Pretoria wanted a more rigorous assessment, hoping this would open up yet more mining opportunities and create a rich stream of revenue. Hall was directed to map out the rocks of the "older formations" and also look for any mineral deposits within them. That was why he set off each day, sitting awkwardly on his horse, with his mapping case, magnifying glass, and compass slung over his shoulders and his geological hammer wedged into a holster. Every evening he returned with a mass of white cotton bags, bulging with angular rock samples and jostling for space, strapped behind the saddle. In this way, he had begun to make sense of the geology of the region: sorting the major types of sedimentary and volcanic rocks, deciding how they might be ordered, laboriously tracing them through the rugged landscape. He labeled the rock formations with names still used by geologists today, taken from nearby localities such as Onverwacht Farm and Moodies Hills, assigning all of them to what he called the "Swaziland System" of rocks—a grand way of saying distinctive rocks found near Swaziland! And he recognized that all these rocks were surrounded by large bodies of once-molten granite. Because Hall did not find any signs of fossilized life, he reasoned that these sedimentary rocks must be some of the most ancient in this part of Africa.

In 1918, Hall was finally ready to publish the results of his work. His most important scientific contribution, however, could be found neatly folded and tucked into a paper pocket at the back of his report. This was a beautifully colored geological map. And on this map was the first tentative image of the rocks I would be studying sixty-two years later. This map did much more than show potential places to find gold. It made geologists throughout the world aware of the rocks in the Makhonjwa Mountains.

IT'S ALL IN THE MAP

Making a geological map would have seemed to Hall an essential part of his survey work. But if you've never made one yourself, the reasons for this might not be obvious. The map is in fact the only practical way to record a vast amount of information about the rocks in a region, guiding all future geological investigation. The rock types are shown in different colors, marking on the map the exact places where they are visible on the ground. Unfortunately, parts of the landscape are also covered by soil, and in those areas the underlying rock must be inferred, or even guessed at, by extrapolating from visible outcrops. So both making and reading a geological map require considerable skill. I have always found the mapping part to be an exhausting, time-consuming undertaking—but extremely satisfying—because it usually involves walking virtually every inch of the landscape, concentrating all the time on marking the rocks on the map and noting down detailed observations about them.

There is much more to the map than a simple record of different types of rock. It needs to help geologists answer many other questions. For example, how are the once-horizontal layers—especially in sedimentary rocks—now oriented due to past shifts in the bedrock? What is the nature of the boundaries between the various rock units? Are they fractures or faults, where the rocks have shifted past each other? Are they sedimentary, where one rock unit has been deposited on top of the other, such as on the bed of a river or at the bottom of a lake or ocean? Or are they intrusive, where one rock has invaded another as a magma? Usually, the answers to these questions pose a whole bunch of new questions! This is why no geological map is ever definitive; it is only after many attempts by different geologists, building on previous work, that the full geological picture begins to emerge. Hall's map of the Makhonjwa Mountains was made over one hundred years ago, and there have been many advances in our ability to read the rocks since then.

The Makhonjwa Mountains are a tiny part of southern Africa, and the bedrock does not just stop at its edges. As with the production of topographic maps, geologists aim for a much more complete coverage with their geological maps, but not necessarily at the same scale: a 1:250,000 map will cover more ground than a 1:50,000 map on the same-sized sheet of paper, but it will show the regional distribution of broad groupings of rocks at the expense of more detailed and local information about individual rock types. When I first started my project, geologists at the British Geological Survey were helping the former Swaziland Geological Survey by putting together a diverse collection of detailed geological maps to produce a large-scale map of their country. To me, who knew nothing about southern Africa, the map that the British geologists were creating certainly seemed very complicated. It compressed together a wide range of geological features, and I remember feeling slightly repelled by the color scheme: garish purples, pinks, blues, greens, reds, and yellows. Comparable maps also existed for the adjacent regions in South Africa, made by geologists from the South African Geological Survey and various universities. And all these maps provided the most up-to-date guide to the bedrock in this part of southern Africa, making it possible to see how the rocks in the Makhonjwa Mountains fit into the much bigger picture of the bedrock of an ancient continent. Let me now describe their main features.

UNCONFORMITIES

I will start with what geologists call an unconformity. Imagine the following sequence of geological events. As part of plate tectonics, two continents move toward each other, closing up an ocean and eventually colliding and pushing up a mountain range. This way, they become welded together to form an even larger continent—a supercontinent. But the new mountains will sow the seeds of their own destruction, because rivers will flow off them, eating into the bedrock and ultimately smoothing out the landscape. At some later stage, the motion of the tectonic plates might change, and the new supercontinent will begin to be pulled apart as the plates move away from each other. Where this happens, the surface will slowly subside. Rivers will flow in, dumping their load of rock fragments carried down from highlands farther away. All this will create a thick pile of sediment, settling in places on the original bedrock of the supercontinent. Geologists call this sedimentary capping the "cover" and the underlying bedrock the "basement." The contact between the two is known as an unconformity.

Unconformities, in my view, are one of the wonders of geology. It is not necessary for continents to collide to create them. Any shift in the landscape that triggers the forces of erosion will lead to one. But where they are visible, you can literally put your finger on a gap in time—the missing geological record has been carried away by the flow of wind, water or ice. Looking at the geological maps of Swaziland and adjacent regions, I saw obvious unconformities where much younger cover rests on an ancient basement collectively known to geologists as the Kaapvaal Craton (see chapter 1). Much of this cover was laid down during the age of the dinosaurs, in the Mesozoic Era. In the eastern part of Swaziland, it mainly consists of a thick pile of volcanic rocks that erupted when Antarctica, Madagascar and India all split off from Africa, as the supercontinent of Gondwana began to fragment into smaller continents. This was the beginnings of India's long journey northwards to eventually collide with Asia about 50 million years ago, pushing up the Himalayas. Compared to the period of geological time I wanted to investigate, these events happened far in the future. My interest lay in those places where the cover had been worn away by erosion, once more exposing the basement and revealing the backbone of the Kaapvaal Craton.

GRANITES AND GREENSTONES

It is the rocks in the basement that contain a record of the early years of our planet. I will focus on just two distinctive groupings of rocks in this basement, very loosely called the "granites" and the "greenstones." I will begin with the granites. These were once molten, intruding deep within the Earth, but they have since cooled and crystallized as a beautiful

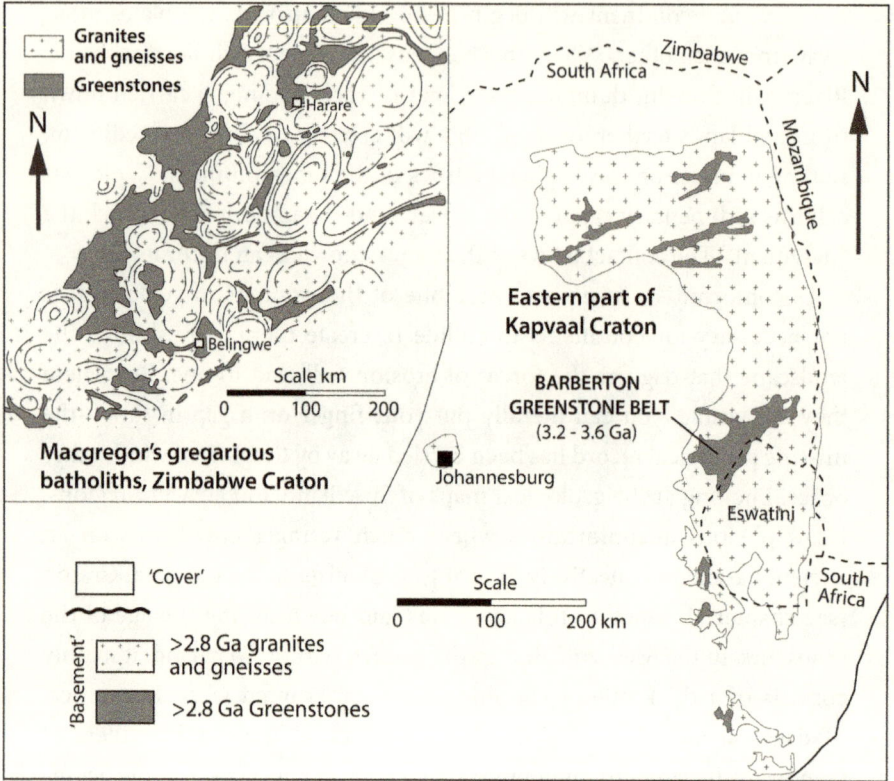

FIGURE 2.2 Parts of the bedrock in southern Africa consist of Archean (2.5 to 4.0 billion years [Ga] old) "granites," with intervening similarly aged wisp-like regions of sedimentary and volcanic rocks called greenstones. These are the ancient cores of the continent—referred to by geologists as the Kaapvaal and Zimbabwe cratons—forming a "basement" on which a "cover" of younger sedimentary rocks was deposited much later on in the geological history of this region. Alexander Macgregor was the first to comment on the remarkable elliptical shapes of the "granites" in the "basement" of the Zimbabwe Craton, calling them "gregarious batholiths" because they seemed to "bubble up" next to each other. (Illustration credit: Simon Lamb)

interlocking mosaic of crystals, composed mainly of quartz and feldspar (though mica, amphibole and garnet can be present too), with individual crystals reaching up to several centimeters (roughly one inch) across. These create a pale-colored speckled rock.

FIGURE 2.3 Metamorphic rock, known as gneiss, on the eastern edge of the Malolotja Nature Reserve, displaying spectacular zebra-like bands of quartz and feldspar (pale layers) and biotite and amphibole (dark layers). These were once a variety of igneous rocks that have been contorted and melted again, deep in the crust, about 3.2–3.3 billion years ago. The pen leaning against the rock face provides scale. (Photo credit: Simon Lamb)

In some cases, especially at the edges of individual granite intrusions, there is a strong flattening of the crystals, giving rise to a tendency for the granite to split easily in particular directions, rather like a stack of cards. Sometimes the granite contains thin layers—centimeters to meters thick—rich in the darker minerals biotite and amphibole, producing a beautiful black-and-white, zebra-like appearance. These form a rock called gneiss (pronounced "nice"), an old German word describing the rock's shiny appearance when freshly broken. Spectacular rock pavements of gneiss are found along many of southern Africa's greatest rivers. The most famous—known as the Sand River Gneiss—are exposed in a tributary of the Limpopo, about 400 kilometers north of Eswatini. Zebra-like pavements of gneiss can also be found in rivers along the edges of the Makhonjwa Mountains. In detail, geologists have established more technical names for

FIGURE 2.4 The Makhonjwa Mountains, in the heart of the Barberton Greenstone Belt, South Africa. In the foreground at left is a ubiquitous protea bush. (Photo credit: Simon Lamb)

the various types of granite and gneiss. In addition to typical granite, there is also tonalite, trondhjemite, and granodiorite, depending on the precise makeup or appearance of the rocks. Geologists like to name a rock after the place where it was first recognized. Thus, tonalite is literally the rock of the Tonale Pass in the Italian Alps, whereas trondhjemite is the rock from Trondheim, Norway. Granodiorite, however, is a name using bits of Latin and Greek, meaning "speckled rock."

In contrast to granite, *greenstone* is a rough-and-ready term often used by miners to refer to a mixed bag of mainly volcanic rocks that can be greenish in color, as well as some sandstones and shales—rocks, in other words, that owe their existence to geological activity on Earth's surface, creating the landscape. Tracts of land between the granite bodies are underlain by these rocks, which early prospectors sought out because they were commonly associated with rich deposits of precious metals, particularly gold. This brings us back to Arthur Hall's pioneering geological map, published in 1918, which I described early on in this chapter. Hall had been instructed to make a preliminary geological survey of the goldfields and volcanic and sedimentary rocks in the Makhonjwa Mountains, extending from northwestern Swaziland (Eswatini today) into the adjacent part of South Africa. He had labeled these rocks collectively on his map as the "Swaziland System," but they also fall exactly into the category of greenstones. Thus, Hall's Swaziland System makes up what geologists call today the Barberton Greenstone Belt (named after the nearby mining town of Barberton in South Africa), surrounded by a bedrock denoted on his map as "older granite." The region we are talking about only covers a small fraction of the Kaapvaal Craton, extending about 110 kilometers at its longest point, and about 50 kilometers at its widest. Its area is roughly that of an average US or English county, or about half the size of the island of Hawaii. But its diminutive size is outweighed many times over, as we shall see, by its geological significance.

The town of Barberton is situated at the western edge of the Barberton Greenstone Belt, on a plain at the foot of the Makhonjwa Mountains. In fact, until recently, geologists talked about the Barberton Mountain Land rather than using the local name Makhonjwa Mountains. Now that I have got my tongue around Makhonjwa, I find I like the African feel it gives to the location of the greenstone belt. Barberton, named after the Barber family, was founded in 1884 after the discovery of a rich gold deposit nearby. Since then, it has been a honeypot for scores of hopeful gold prospectors and once provided all the services for a mining industry that has gone through many booms and busts. Percy FitzPatrick wrote about his time in Barberton in the classic children's book *Jock of the Bushveld*, published in 1907. A square in the town has a statue of Jock, who (in case you didn't know) was an adventurous dog living amongst miners, cattle drovers, and various dubious local characters.

A GREGARIOUS NATURE

In order to visualize the pattern made by the granites and greenstones on a large-scale geological map, imagine cutting biscuits for baking from a sheet of pastry using a circular cookie cutter. Naturally, you will want to make the best possible use of the pastry, so you cut the biscuits as close together as you can. Once you have cut all the biscuits you will be left with a strange wisp-like network of unused pastry. If you color the biscuits pink, and the unused pastry green, you will have something like the map of the rocks in the basement, where the pink now represents any granite, and the green represents the wide variety of volcanic and sedimentary rocks making up the greenstones. And the same wisp-like greenstones between more circular domains of granite have been recognized not just in the basement rocks of Eswatini and South Africa, but also farther north in Zimbabwe—and in other continents too, such as western Australia and northern Canada. If you know what you are looking for, you can actually see them clearly from space in modern satellite imagery. All of this indicates that the presence of granites and greenstones in the continents is a widespread geological feature of our planet. However, we also now know that there is considerable variation in the details of the rock types present in any particular region; they will have widely different ages and geological stories—not surprising, given that they span nearly two billion years of Earth's early history.

The distinctive shapes of the granites and greenstones in southern Africa were first systematically described by a remarkable geologist in the Rhodesia Geological Survey (as it was then) called Alexander Miers Macgregor. Between the two World Wars, he devoted his life to mapping the rocks in Rhodesia. In 1951 he summarized much of this work in a famous presidential address to the Geological Society of South Africa, in which he coined the wonderful phrase "gregarious batholiths." To a geologist, a batholith is a large body of magma that solidified deep in the crust—from the Ancient Greek *bathos*, meaning "deep," and *lithos*, meaning "rock."

I suppose, in Macgregor's mind, the word "gregarious" conveyed the idea of many batholiths congregating together through a tendency for new ones to rise up close to previous ones, forming a distinctive feature of the ancient rocks in the continents. I wonder if he ever saw the strange coincidence between this word and his name, especially as his idea is usually referred to today as "Macgregor's gregarious batholiths"? Gregarious batholiths have certainly exerted a strange hold over geologists ever since. Somehow, whenever one sees that broad pattern of granite and greenstones on a geological map of southern Africa, there is a tendency to think of it as just the remains of a vast bubbling mass of hot rock, part and parcel of the volcanic activity in the overlying greenstones, although as I will show you in the rest of the book, the truth is far more surprising and fascinating.

There was, however, one immediate consequence. Here was a way to estimate the age of the rocks by dating when some of the surrounding granites intruded into the greenstones. It turns out that the pioneers in this dating were based not far from the Barberton Greenstone Belt, at the Bernard Price Institute of Geophysics in Johannesburg. I am very familiar with this institute, because I visited it many times during my field seasons. I was also privileged to be able to get to know its director at the time: Louis Nicolaysen, one of South Africa's most distinguished scientists. In the early 1960s, he played an important role in developing an ingenious new way to use natural radioactivity to date rocks, laying the foundations of modern geochronology—literally, the study of Earth time.

TELLING THE TIME

I have already introduced the idea in the previous chapter that natural radioactivity can act as a clock in rocks. Making use of this clock, however, involves delving into the very nature of matter and requires a special (and very expensive) measuring tool called a mass spectrometer. The basic principles had been worked out in the early 1900s by Ernest Rutherford and Frederick Soddy, when they and others discovered that there are different versions of elements called isotopes. All isotopes of a particular element are virtually indistinguishable to a chemist, behaving in much the same way in a chemical reaction. But to a nuclear physicist they are quite different, because they vary in atomic mass. Some isotopes are unstable and undergo radioactive decay, transforming into different elements and emitting potentially hazardous radiation in the process. In other words, a parent radioactive element decays into a daughter product. It happens at a precise rate, characterized by the time it takes for a population of radioactive atoms to decrease by half. This is known as the half-life, and it can be very accurately measured in the laboratory. Crucially, the half-life is not affected by temperature, pressure, or any known chemical reaction, and so regardless of how a rock formed, the radioactive clock keeps the same time.

Yet there was a problem with the clock, as Louis Nicolaysen and his colleague Hugh Allsopp came to realize. Just like resetting a stopwatch, a datable geological event needs to somehow set the radioactive clock to zero, and it was not always clear how this would happen. The age of an igneous rock is taken to be the time elapsed since it cooled from a magma. Geologists sometimes simply assumed that as the magma cooled, it would crystallize as minerals that only contained the parent radioactive elements. This way, the radioactive clock would be set to zero at the moment of crystallization, and the subsequent accumulation of daughter product would be a measure of the time since the rock solidified. This works well if the daughter product is a gas that can escape easily from the magma but is trapped in the solid rock. I have routinely used measurements of the decay of radioactive potassium into argon to date igneous rocks, as part of my geological research in the high Andes of Bolivia. Argon is what is known as a noble gas, which means that it does not usually react with other elements but instead leaks out of the magma. The magma will be virtually free of argon when it cools and begins to crystallize as a solid rock, but as the radioactive potassium decays, argon will now build up in the rock. Louis and Hugh realized that in most decay systems, significant amounts of the daughter product will also be trapped in the magma, preventing the resetting of the radioactive clock. This will result in a completely spurious age for the rock.

Geologists needed a much more versatile way to date their rocks. Louis and Hugh worked out a solution: the "isochron" method, which is now the standard method of dating. I eventually met Louis in 1981, at the start of my first major field season working in Swaziland, and he took great interest in my work. He was a very courteous man with a sort of half-smile on his face. I was introduced by Maarten de Wit, a giant Dutchman who will play a prominent role later in this book. Maarten was also working at Louis's institute, and he kept reminding me with an expression of awe that "Louis was one of the people who invented the isochron." The method turns out to be rather complicated to explain, so I will only give you a flavor of how it works. Louis and Hugh wanted to use the commonly occurring radioactive parent–daughter elements rubidium and strontium, neither of which are gases, to date the granites in southern Africa. This seemed a good method for dating granite because there was enough of these elements to be easily measured with the techniques available at the time.

Louis and Hugh's brain wave was to take advantage of the fact that not all strontium is produced by radioactive decay of rubidium. There are other isotopes that remain unchanged over time, and these can be identified in a mass spectrometer. This way, Louis and Hugh had, in effect, found two clocks in one. One clock measured the decay of rubidium to strontium, and the other measured relative changes in the strontium isotopes. Since the team knew that different types of minerals vary in how much rubidium they contain, they analyzed these minerals individually. Crucially, because the minerals had all crystallized from the same body of magma, they assumed that each one began with the same proportions of the strontium isotopes. Louis and Hugh showed that it was now possible to calculate a unique age for all the minerals in the rock—this is the "isochron" or "same age."

Louis and Hugh set up a laboratory at the Bernard Price Institute; it was dedicated to dating a wide range of rocks in South Africa with this new approach to reading the radioactive clock. By the mid-1960s, this work had changed our conception of Earth's history. Previously, the oldest widely accepted ages for rocks were less than 3 billion years. The new results showed that the oldest "granites" near the edges of the Barberton Greenstone Belt were much older than this—over 3.4 billion years old—which meant that at least some of the rocks that make up the Barberton Greenstone Belt must be even older. At the time, these were the oldest known rocks on Earth. But there were also granites that were several hundred million years younger. These results demonstrated the huge amounts of geological time represented by the ancient rocks in the continents.

A MICROSCOPIC TIME CAPSULE

By the time I started my study of the rocks in the Makhonjwa Mountains, the general sequence of geological events had already been revealed from the ages of the surrounding granites, dated using Louis and Hugh's isochron method. However, in the last thirty-five years or so, geochronologists such as Alfred Kroner at the Max Planck Institute in Mainz, Germany, have discovered a sort of silver bullet that has made it possible to accurately date a much larger variety of these very old rocks, in particular those making up the Barberton Greenstone Belt itself. This is a tiny crystal of zirconium silicate called zircon, whose age can be determined by measuring the decay of radioactive uranium to lead. Zircon crystallizes at relatively high temperatures from the magma, and when this happens it incorporates small amounts of radioactive uranium—but crucially, virtually none of the daughter product lead—into itself. As a result, the uranium-lead radioactive clock is essentially set to zero in every zircon crystal. Thereafter, the crystal is almost immune to the subsequent history of the rock, providing a time capsule that holds the key to when the rock solidified. And the dates obtained this way are incredibly precise, although they require the use of a microscopic laser beam to probe the crystal interior. Thus, zircon naturally solves the problem Louis and Hugh had recognized when they came up with their isochron.

A New Zealand colleague of mine, Cornel de Ronde, was one of the early geologists to make sense of the new zircon ages for the Barberton Greenstone Belt. Working together with Maarten de Wit, he published a detailed timetable for its formation in 1994. Since this pioneering work, the timetable has been refined by the addition of many more zircon dates, although there are still large gaps in time spanning tens of millions of years or more for which the rocks remain undated. You can get a better feel for the precision of these ages if I quote them here in millions of years instead of billions. The oldest ages of all come from granites that form part of gneiss in Eswatini, dated to 3,644 million years ago, plus or minus 4 million years. The oldest dated volcanic rocks in the Barberton Greenstone Belt have an age of 3,538 million years old with an uncertainty of plus or minus a few million years—this is like estimating the age of a middle-aged person to the nearest week or so. And the youngest volcanic rocks in the greenstone belt are 3,216 million plus or minus 0.3 million years old. The closing act in this story was the intrusion of large bodies of granite—the big "biscuits" in the cookie-cutter analogy—which crystallized between 3,227 and 3,140 million years ago. These terminated the geological life of the Barberton Greenstone Belt and marked the birth of the Kaapvaal Craton. Together, the granites and greenstones record part of the adolescence of the Earth over a period of about 500 million years. And like a young person, the surface of the young planet seemed unable to sit still.

It is probably wise for me to declare that the granites and greenstones of the Kaapvaal Craton, although they are some of the oldest rocks so far found in Africa, are no longer the oldest known on the planet. Finding these has become a source of intense rivalry between research groups, with a paper in a prestigious scientific journal such as *Nature* or *Science* as the prize. For many years, I worked in the same university department at Oxford University as Stephen Moorbath, who was a pioneer of this research. I took over Stephen's office when he retired, and I inherited his old filing cabinets, labelled with the places in the world where he had dated the rocks. Back in the 1970s, Stephen broke all records when he showed that the volcanic and sedimentary rocks exposed at Isua, right on the edge of the Greenland Ice Sheet, are 3.7 to 3.8 billion years old.

The Isua rocks are only visible in two tiny regions, each about 2 kilometers wide and 15 kilometers long, and they have been both cooked up and squeezed by the huge pressures in the Earth. Nonetheless, they are a unique window into the early history of our planet—I will come back to them later in this book. Intriguingly, as the Greenland Ice Sheet retreats further during global warming, more of the Isua rocks will be revealed, and there will be new discoveries about their significance. Certainly, more outcrops of very ancient rocks remain to be discovered in the cratons—geologists are already arguing over a number of possible candidates.

Sam Bowring at the Massachusetts Institute of Technology broke the "four billion years" barrier when he dated zircon crystals in the so-called Acasta Gneiss in northwestern Canada. Even older ages have been obtained for zircon crystals found in Western Australia. These only exist as grains of sand in much younger sedimentary rocks, plucked from a more ancient bedrock by the forces of erosion. They prove, however, that there are relics of early Earth as much as 4.4 billion years old. This is in the earliest part of Earth's history, called the Hadean Eon after the mythical ancient Greek underworld of Hades, and it was once considered a time when the surface was mainly a molten inferno, beyond the reach of geologists. As I have already described, it is after 4 billion years ago that

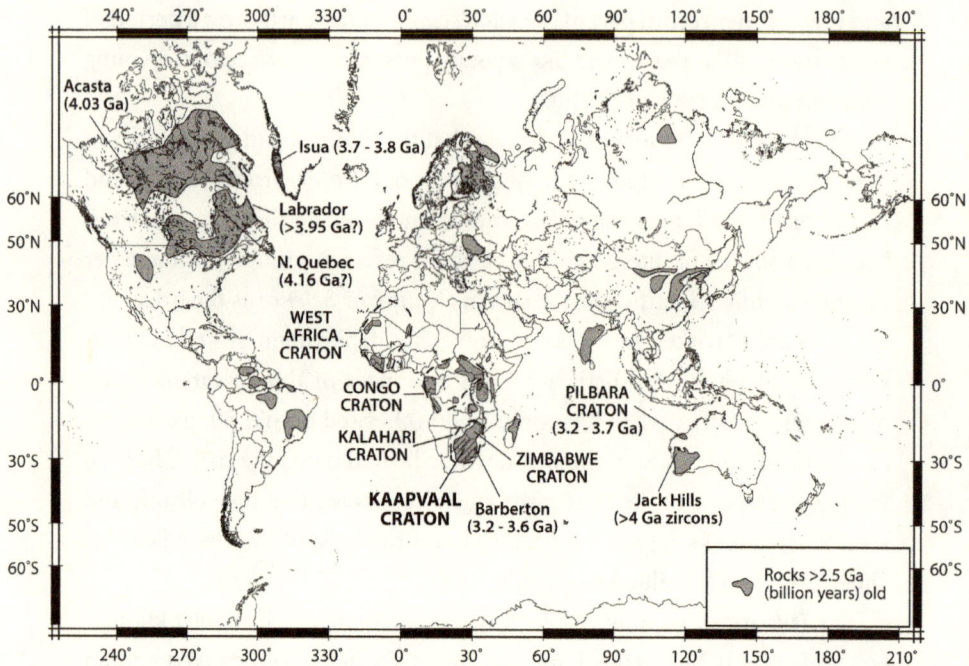

FIGURE 2.5 The oldest rocks on Earth and where to find them. Rocks that formed over 2.5 billion years ago, during the Archean and Hadean eons, are confined to relatively small areas in the ancient cores of the continents. Those in the Kaapvaal and Pilbara cratons preserve an unusually rich record of Earth's surface conditions from 3.2 to 3.6 billion years ago, while the ancient rocks at Isua in Greenland, and also in Canada's Northern Quebec and Labrador coastal region, offer rare glimpses of an even older surface. The oldest relics of Earth's crust are zircon crystals preserved among grains of sand in younger sedimentary rocks, particularly abundant at Jack Hills in western Australia. (Illustration credit: Simon Lamb)

the rock record really begins—in the succeeding Archean Eon, derived from the Ancient Greek *arkhē*, meaning beginning or origin. However, it is not until about half a billion years into the Archean that the record becomes prolific, around 3.5 billion years ago—in other words, toward the end of the first quarter of Earth's history. It is now preserved in the rocks of the Barberton Greenstone Belt and also the Pilbara region of Western Australia.

So, what had happened during these early "years" of our planet's history? My task was to find some answers to this question using the evidence from the Barberton Greenstone Belt where it forms the bedrock of the Makhonjwa Mountains in Swaziland (now Eswatini). I clearly needed to go there and have a look for myself. Thus it was that I found myself in July 1980 stepping off the daily Royal Swazi flight from Johannesburg, at the small airport of Manzini.

- § -

IN THE FIELD

The portion of the Makhonjwa Mountains that lies within Eswatini is a region of special scenic beauty and home to a wide range of endangered plant and animal wildlife. But back in the 1970s, local small holders were encroaching on the mountains, and the slopes were beginning to suffer the effects of overgrazing from the large cattle herds that are the traditional form of Swazi wealth. The desirable cycad plants were also being illegally dug up and sold to collectors. In 1979, the King of Swaziland was persuaded to designate a strip of land (about 7 kilometers wide and 27 kilometers long) along the international border with South Africa as a national nature reserve. The reserve was called the Malolotja (pronounced and sometimes spelled Malolotsha) Nature Reserve, after the Malolotja River Valley that lies at its heart. The Malolotja River is a short tributary of the Komati River, the latter rising much farther west in South Africa before flowing across the reserve, and then on through Eswatini and Mozambique to the sea. The Komati River is famous to geologists throughout the world, because, as I will explain in chapter 3, it has given its name to a rock that except for some very rare examples, is only found amongst remnants of early Earth.

I had already decided back in England, after looking at the geological map of Swaziland, that the rocks I needed to study lay inside the reserve. The reserve was managed by the Swaziland (now Eswatini) National Trust Commission, or SNTC for short. For this reason, soon after I arrived in Swaziland, I arranged through the Geological Survey to meet the head of the SNTC, who turned out to be a white Swazi—a settler of European origin who had decided after independence from the British in 1968 to commit to Swaziland rather than Britain. We met in his office: an inexperienced student of twenty-one, looking across the desk at a seasoned administrator and zoologist dressed in loose khaki safari clothes. His concern was that I would disturb the wildlife in the reserve and also become a liability if I ran into trouble by getting lost or falling over a cliff (little did he know about my future car accident, although that was not in the reserve). However, he could clearly see that by carrying out basic scientific research, I was fulfilling one of the important aims of the SNTC. I got my permission, enshrined in an official letter. But I was told to strictly follow the instructions of the warden.

The outstanding problem was working out how I could actually get into the reserve. The warden and rangers used Land Rovers provided by the SNTC, but there was no question of me having access to one of these. I was left scrambling for a solution, including privately hiring a car and later purchasing a succession of vehicles. None of this was straightforward, and as you will see, I came up with a variety of solutions which all had their own problems. Organizing a means of transport took time and energy, causing me a considerable amount of anxiety. This meant that when I had finally found a way to reach the rocks in the reserve—going into the field, as geologists say—I had not thought as much, as perhaps I should have done, about what I was going to do once I got there. The rocks I was about to be confronted with have proved to be some of the most challenging that any geologist—especially an inexperienced one like myself—could possibly deal with. There were some particular aspects of the rocks that I needed to understand if I was to have any chance of success with my project. So, before I tell you about my own experiences, I will introduce you to what they were.

FINDING THE RIGHT ORDER

Imagine a catastrophic volcanic explosion that creates a vast cloud of dust. This will eventually rain down on the ground, forming a layer of volcanic ash. Not long after, a rent opens up; lava pours out and covers the volcanic ash. Subsequently, the volcano becomes quiescent, and the lava flow is buried by deposits of sand and silt from the flooding of a nearby river, building up yet more horizontal layers one on top of the other. Much later, if a river cuts through the layers to carve out a steep-sided gorge, there will be distinctive horizontal bands displayed in its wall. Now, along comes a geologist who wants to read the rock record here! They would find working out the order of the layers a trivial matter, because common sense says that the ones on top are younger than those at the bottom. But the bedrock does not always present a geologist with such a simple situation. Consider what would happen to the layers if, after the geological events I have described, the region also became caught up between two colliding tectonic plates.

The colliding plates would bring huge horizontal forces to bear inside Earth. These forces are capable of tilting layers, turning them on end so that they resemble a row of books on a bookshelf. Now their order is no longer obvious, because one can read such a "bookshelf" of layers in two ways: either from right to left, or left to right, in order of age. In the absence of extremely precise ages for the rocks, our geologist could plausibly read the sequence of layers as a volcanic explosion followed by a lava outpouring and then burial during river flooding—or completely the other way around. This confusion might not seem particularly important on this small local scale, for such a short period of geological history. But it becomes very significant if one is trying to read a record in sedimentary rocks involving many hundreds or thousands of layers, especially during the early history of Earth when there are no fossils to help with the ordering. Fortunately, geologists have other tricks up their sleeves.

If you listen to a group of geologists out in the field, scrutinizing a stack of layers, you may well hear the question "Which way up?" repeated many times, followed by either an argument, or somebody bending down and putting their face close to the rock surface. The geologists are trying to determine the original tops of the individual layers—or bedding, as they tend to call it—which indicate the direction that was originally up (facing the sky) when the beds were deposited. This, then, uniquely defines their local order from oldest to youngest. The trick is to look at the details of how the individual layers formed. Looking closely at sandstones has proved to be the key to this.

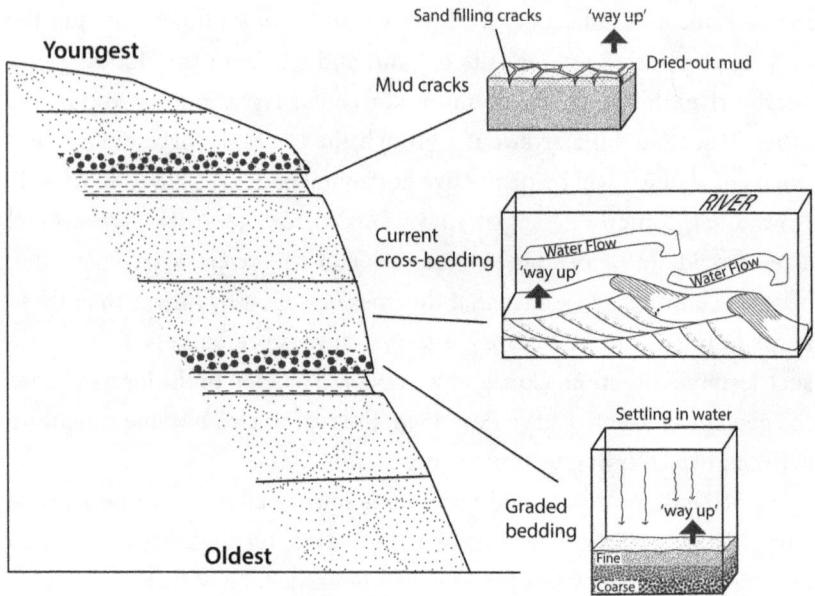

FIGURE 2.6 Which way is up? An essential part of studying sedimentary rock sequences is determining the order of the strata from oldest to youngest: what geologists refer to as the "way up." This information is revealed through detailed features in the sedimentary layers. (Illustration credit: Simon Lamb)

In a desert, turbulent gusts of wind whip grains of sand into the air and then dump them back on the ground. These apparently random movements of sand are creating something of immense beauty and regularity in the desert landscape: sand dunes. Dunes are formed in much the same way underwater—on the beds of rivers, lakes or seas—and range in size from much smaller than your hand (called ripples) to the size of a large building. Instead of the wind, it is the current of water that moves the sand along. In time, the dunes will be preserved in the sand layers deposited on the riverbanks, and as the layers are buried more deeply, they will become features of solid rock, fossilized in sandstones.

The inside of the dune or ripple is built up of numerous layers set at angles to one another, with some truncating others, created as the sand reshapes itself in the current. The whole effect is called cross-bedding.

These are preserved as thin laminations picked out by grains with slightly different sizes or colors, and if you look at broken surfaces of the rock with the practiced eye of a geologist, they can be "read." This way, depending on the details of the cross-bedding, it is possible to work out which way the river flowed, or in the case of desert sand dunes, the prevailing wind direction. Importantly, given our present discussion, the way the internal laminations cut across each other reveal which way is up—the younger laminations cut across older ones like mini unconformities. There are many other features that geologists also look for. There may be the remains of muddy layers that cracked when the riverbed briefly dried out, with sand then filling up the cracks. Or the grains of sand may show a characteristic grading because the larger grains have settled out first, followed by progressively smaller ones. Again, the practiced eye of a geologist can tell from all this which way is up, and thus the local order of the layers.

I could go on with yet more nitty-gritty detail about the layers of sedimentary rock. But by now, you should have a good idea of the sorts of things that occupy a field geologist's mind as they try to make sense of the layers and extract as much information as possible about the history of the rocks. These are concepts I had learned about in my undergraduate course work, although more in theory than practice. Somehow, I had to now apply them in my new field area to rocks I had never seen before. It was obvious to me that I still needed help from somebody who was already familiar with the rocks in the Barberton Greenstone Belt. Also, as a requirement of my postgraduate degree, it was not enough to just describe what I had found myself; I had to demonstrate that I was significantly advancing geological understanding with new discoveries and analyses, building on the large body of previous work in this field of research.

If all this was not daunting enough, I needed to regain my confidence in working in remote parts of southern Africa after the terrible accident I described in the previous chapter, when my vehicle had rolled off the side of a mountain and plunged 50 meters (about 150 feet) into a deep gulley. It turned out that my unexpected meeting with the Dutch geologist Maarten de Wit helped me achieve all these goals in one fell swoop.

MEETING MAARTEN

The accident completely overwhelmed me. I had very nearly been killed, and the only thing now, I thought, was to get home, back to England, as soon as possible. I remember going to the Geological Survey offices in Mbabane, the main town in Eswatini, to let them know. But to my surprise I was handed a piece of paper with a handwritten telephone message. Maarten de Wit, at the Bernard Price Institute of Geophysics in Johannesburg, was also investigating the Barberton Greenstone Belt in the Makhonjwa Mountains on the South African side, and he had heard about my project. He was going to be in the town of Barberton, staying at the old hotel. If I could get myself there in a week's time, then he would show me his new discoveries about the rocks.

Coming at this time, Maarten's message made me realize that I was not alone. My whole mood changed instantly. Somehow, I had to find a way to get to Barberton on time. This involved entering South Africa and traversing the Makhonjwa Mountains on a rough dirt road. There were no buses. And I had no transport of my own. Faced with this problem, I became the proud owner of the first trail motorbike to be sold in the country—you can learn at the end of this chapter about my triumph over adversity to achieve this. A few days later I set off to Barberton, heading northwest from Mbabane, skirting the Malolotja Nature Reserve before descending into the deep Komati valley on a road with endless hairpin bends, then climbing all the way up the other side. I turned off onto a potholed and orangish-brown dirt road that took me to the border crossing at Bulembu, perched on a ridge and enveloped in clouds. From Bulembu, the dirt road continued, winding its way across the Barberton Greenstone Belt before finally descending into Barberton itself. Covered in road dust, I had just ridden over much of what is left of very early Earth in this part of Africa.

Maarten de Wit met me with a huge smile on his face, and somehow all my worries seemed to fall away. He was thrilled by my motorbike. With him was a young French student, Isabelle Paris, who had been sponsored by a mining company to undertake a similar project to mine, but within South Africa. Maarten clearly wanted us to work together. It was during this trip that I started learning how to recognize the many peculiar rocks in the Makhonjwa Mountains. I certainly began to appreciate the importance of knowing which way is up for the sedimentary layers. But what made the experience particularly eye-opening was the contrast between the traditional way of thinking about these rocks, and Maarten's new ideas.

RECEIVED WISDOM

A group of geologists out in the field can sometimes display traits of an aggressive lawyer with a penchant for harshly interrogating suspects in an attempt to undermine their alibis, but for geologists the suspects are inanimate rocks and the alibis are previous geological ideas.

We were no different on our field trip through the Makhonjwa Mountains, questioning almost everything about the underlying bedrock. Nonetheless, I was in awe of this bedrock because I knew enough to be excited by the fact that this was the world-famous (at least to geologists!) Barberton Greenstone Belt, and I was standing right in the middle of it. I desperately wanted to understand not only what Maarten had come up with, but also the previous way of thinking about these rocks, or what I now thought of as the "standard model" (adopting the terminology of particle physicists). Much of the standard model had been developed from mapping done by geologists in South African universities and the Geological Survey from the 1940s through the 1970s. Some of the big names from this period were Morris and Richard Viljoen, and Carl Anhaeusser. Carl has continued his research up to the present day; he is a highly respected geologist with vast experience who is still publishing scientific papers on aspects of the geology of these rocks.

When Maarten had first started his study of the Barberton Greenstone Belt, it was a commonly held view that there was not much new to discover. The sedimentary and volcanic rocks had already been assigned to three major groupings, lying one on top of the other. Some geologists had even gone so far as to identify equivalent groupings in other continents too. All this was part of Arthur Hall's "Swaziland System," first tentatively shown on his 1918 map. The standard model still used some of his original names for the groupings while adding many more subdivisions, all taken from the local farms or places where the particular rocks were found. So, there was the Moodies Group that formed the bedrock in the Moodies Hills, and the

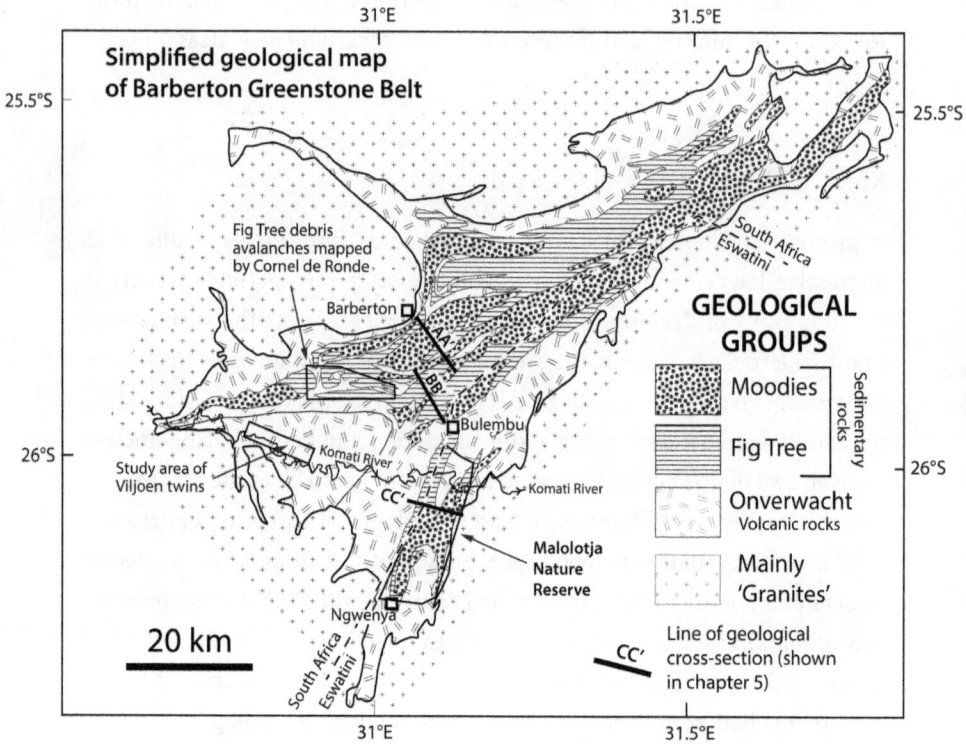

Simplified geological map of Barberton Greenstone Belt

31°E 31.5°E
25.5°S 25.5°S

Fig Tree debris avalanches mapped by Cornel de Ronde

Barberton

South Africa / Eswatini

GEOLOGICAL GROUPS

Moodies — Sedimentary rocks

Fig Tree

Onverwacht — Volcanic rocks

Mainly 'Granites'

Bulembu

26°S 26°S

Study area of Viljoen twins

Komati River

Komati River

CC'

Malolotja Nature Reserve

Line of geological cross-section (shown in chapter 5)

20 km

Ngwenya

South Africa / Eswatini

31°E 31.5°E

FIGURE 2.7 A geological map of the Makhonjwa Mountains, which straddle the border between South Africa and Eswatini. The bedrock, surrounded by "granites," is made up of one of the oldest known sequences of volcanic and sedimentary rocks (dated to 3.2–3.6 billion years ago). Geologists call it the Barberton Greenstone Belt, after the nearby town of Barberton, and its distinctive "wispy" shape is due to the distortion of the rock layers by the surrounding "granites" that have forced their way up through the crust. The volcanic and sedimentary rocks are divided into three distinctive groups—see the next figure for more detail; profiles of sideways views (geological cross-sections) of the rock layers are given in chapter 5. Over the years, there have been many detailed studies of these rocks. My own work was mainly in the Malolotja Nature Reserve, where the Komati River has cut deeply into the bedrock on its way eastward toward the Indian Ocean. (Illustration credit: Simon Lamb)

Generalised sequence of rocks
in the Barberton Greenstone Belt

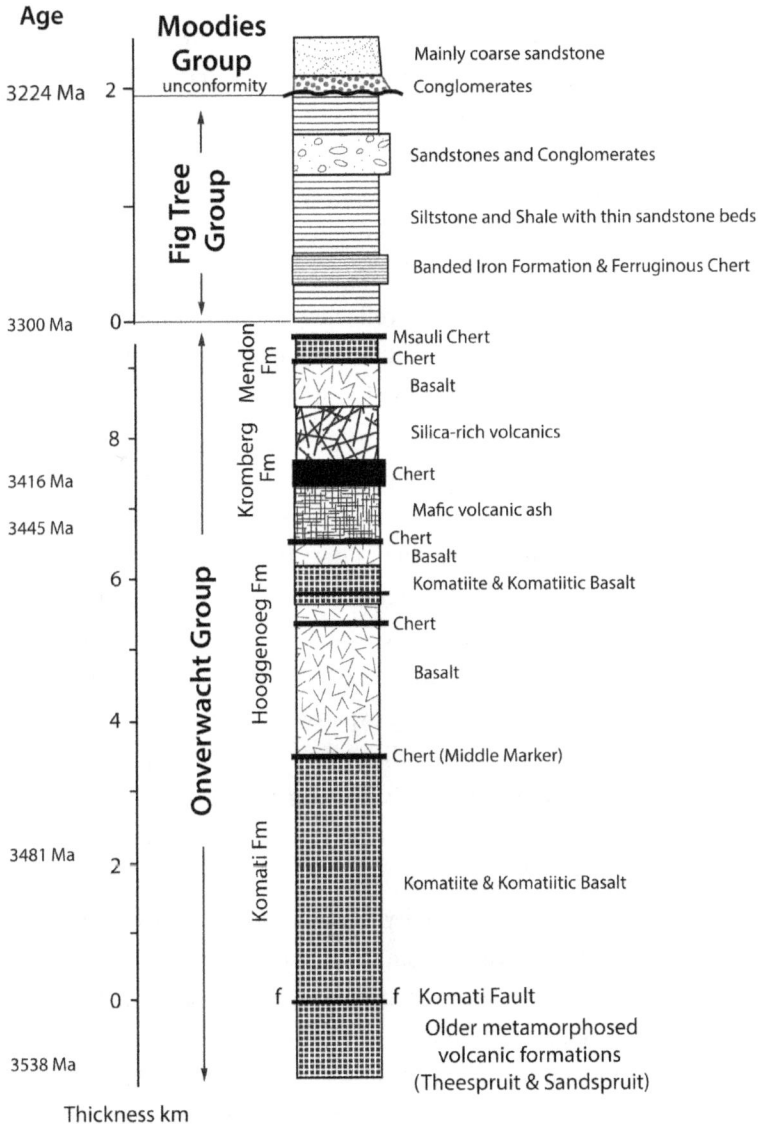

Age					
3224 Ma	2	Moodies Group	unconformity		Mainly coarse sandstone
					Conglomerates
		Fig Tree Group			Sandstones and Conglomerates
					Siltstone and Shale with thin sandstone beds
					Banded Iron Formation & Ferruginous Chert
3300 Ma	0				

Onverwacht Group

Mendon Fm — Msauli Chert / Chert / Basalt

Kromberg Fm — Silica-rich volcanics / Chert / Mafic volcanic ash / Chert / Basalt

8

3416 Ma

3445 Ma

Hooggenoeg Fm — Komatiite & Komatiitic Basalt / Chert / Basalt

6

Chert (Middle Marker)

4

Komati Fm — Komatiite & Komatiitic Basalt

3481 Ma   2

0   f ———— f   Komati Fault

Older metamorphosed
volcanic formations
(Theespruit & Sandspruit)

3538 Ma

Thickness km

FIGURE 2.8 The sequence of rocks in the Barberton Greenstone Belt records over 300 million years of Earth's early history, from about 3,540 to 3,220 million years ago. It begins with a range of unusual volcanic rocks and cherts that formed an ancient ocean floor, known as the Onverwacht Group. This is overlain by sandstone and siltstones of the Fig Tree Group, deposited at the edges of this ocean, both along the shores of volcanic islands, and in much deeper water farther offshore. Finally, coarse sandstones and conglomerates of the Moodies Group were laid down by rivers and near-shore currents. (Illustration credit: Simon Lamb)

Fig Tree Group in Fig Tree Creek (where there was perhaps a wild fig tree?), and the Onverwacht Group after a farm called "Onverwacht."

The rock names that intrigued me most were those clearly linked to Boer settlers, such as the Onverwacht Group and some of its subdivisions—for example, the Hooggenoeg and Kromberg Formations, also named after local farms. I asked Maarten about the meaning of these Dutch-sounding names. He told me that in Afrikaans, *onverwacht* means "unexpected," and *hoeggenoeg* and *kromberg* mean "high enough," and "black mountain" respectively. Perhaps these names reflected the early experiences of the Boer settlers. I have this vision of Boer trekkers rounding a corner to be surprised by the landscape ahead, with a gloomy hill in the distance, later calling a halt after their wagons had struggled up some steep hillside!

I find it easiest to explain the standard model by describing it in terms of the widely assumed geological stories for the various groupings of rocks. I have already talked about attempts to date these ancient rocks, so I will just focus on their interpretation in terms of the landscapes at this time. The Onverwacht Group was thought to be a thick pile of volcanic rocks erupted from volcanoes on land or in shallow water—perhaps a monotonous landscape of volcanoes with intervening expanses of water, looking like parts of Iceland? But slowly, over time, the nature of the volcanoes seemed to change. They became more explosive, spewing out the sort of lavas that one might find coming from volcanoes in the continents, perhaps outpourings from deep bodies of molten granite.

Eventually, this volcanic activity ceased and the volcanoes became extinct. The volcanic landscape began to subside, creating a shallow sea that filled up with fragments of rock washed in by rivers from surrounding regions. A relatively featureless landscape a bit like Hudson Bay? This way, the Fig Tree Group accumulated on top of the Onverwacht Group. Even later still, those gregarious batholiths started to become restless. The rivers continued to wear away the landscape, exposing the batholiths' deep roots. Sand carried away from the granite hills built up the beds of the rivers, now preserved as the Moodies Group, perhaps creating a landscape like that of Mongolia or other parts of central Asia, although there seems to be evidence for coastlines too?

This vision of early Earth was really a leftover of the way geologists interpreted sequences of sedimentary rocks before they had the paradigm of plate

FIGURE 2.9 Geologists enjoying a break. Isabelle Paris (left), Maarten de Wit (center), and me, photographed with my camera's self-timer during my first trip to the Barberton Greenstone Belt. (Photo credit: Simon Lamb)

tectonics to guide them. There was no fundamental explanation for why things evolved the way they did, and as Maarten subsequently discovered, the standard model had taken the layers too much on trust, so to speak, by assuming they were all in the right order with the same upwards direction. Maarten pointed out another puzzling feature: the standard model resulted in a pile of volcanic and sedimentary rocks that reached nearly 20 kilometers in thickness. In other words, when rivers were depositing the last of the Moodies Group, the layers in the oldest parts of the Onverwacht Group would be nearly 20 kilometers below the surface, deep in the crust. The temperatures in this pile would be expected to rise as the rocks became more deeply buried, becoming high enough for new minerals to crystallize, flattened by the colossal weight of the overlying rocks. Although the rocks were clearly "cooked up" where they had come into contact with once-molten granite, there were no signs of this away from the granites, with almost perfect preservation of many features of the rocks deep inside the pile.

Maarten had spent the last two years examining the rocks in the Barberton Greenstone Belt, and he had come up with a new vision of the surface of early Earth as a dynamic world that was constantly renewing itself, pulled and pushed by the forces of the earliest plate tectonics. This was the vision he unfolded to us. Fundamental to his interpretation was a re-evaluation of which way was up for the layers of rock. By focusing on this, using all the tricks I mentioned earlier, he saw the rock strata in a new light, as I shall describe, repeatedly twisted back on themselves by giant folds. Maarten showed us a topsy-turvy world where nothing was quite what it seemed.

A REVELATION

As we bumped along in Isabelle's ancient Land Rover, lent to her by her sponsoring mining company, Maarten pointed out the ridges of rock exposed in the distance. We were cautiously making our way along an access track, bulldozed into the steep hillside so that electricity workers could maintain the cables and pylons that traversed this remote region. Maarten stopped the Land Rover where the track crossed one of these ridges. It was a layer of chert, just like flint, smashing into myriad razor-sharp shards if hit with a hammer—almost a pure silica rock. Yet, if one looked closely, it was full of beautiful spheres, outlined by slightly whiter chert against a darker background. The spheres were laid out in thin sheets, sometimes showing grading, sometimes arranged in wavy laminations—like warped plywood viewed end-on—or even displaying cross-bedding. These were the unmistakable features of ripples and dunes. We were looking at something that had been carried by the flow water, possibly balls of volcanic ash that had clumped together around raindrops during an explosive volcanic eruption, creating what are called accretionary lapilli. Accretionary lapilli are commonly found in modern eruptions; they grow like miniature snowballs in the ash cloud before raining down on the ground. In water, they can be swiftly carried away and deposited with all the features of the ripples and dunes, such as the cross-bedding preserved in sandstones.

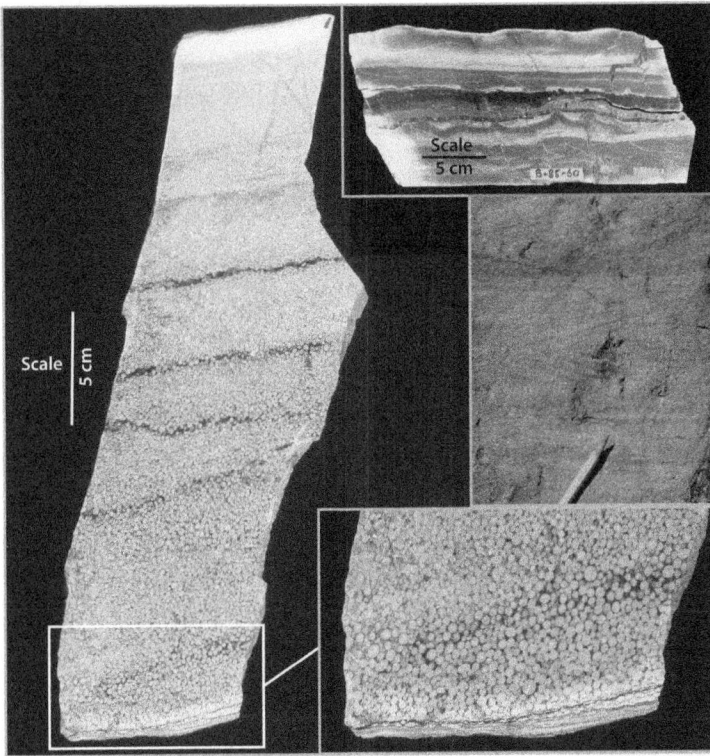

FIGURE 2.10 Chert, chert everywhere! A common rock in the Barberton Greenstone Belt is chert, composed primarily of silica (silicon dioxide). These cherts were originally a variety of other rocks, now radically altered in chemical composition, though they retain enough clues to reveal their origin. One common type formed from balls of volcanic ash that clumped together in volcanic eruptions as accretionary lapilli. These were subsequently carried and sorted by underwater currents, accumulating as graded beds typical of deep water turbidites where larger lapilli settle at the base (*bottom right and left photographs*: courtesy of Cornel de Ronde), and smaller ones higher in the layer. Cross-bedding and ripples are also visible (*top photograph*: courtesy of Cornel de Ronde; *middle right*: photo by Simon Lamb).

We found other chert layers. But this time they had a pale green color, with very different internal features. Greenish chert patches perfectly mimicked the crystal shapes typically found in rocks that had crystallized from a molten state. This had all the characteristics of a lava, yet it was, once again, pure silica—like flint. Elsewhere, Maarten showed us cherts that must have once been conglomerates, containing the ghostly outlines of multicolored fragments, several centimeters across, of white, black, red, grey, and green chert. It was becoming clear that the rocks here were in disguise. They had originally been produced by a wide range of geological activity—volcanic eruptions, ash clouds, underwater currents, and erosive forces—but now they all looked the same from a distance, fooling geologists.

Maarten gave us his explanation. Intense volcanic activity must be constantly flushing hot water through cracks in the bedrock, and a powerful cocktail of elements in this water—particularly silica and iron—was reacting chemically with the rock, drastically changing its composition. Although this often obliterated any features of the rock, sometimes—and rather mysteriously—it preserved them in extraordinary detail. There was a dividend to all this, because these features could be used to determine which way was up. It became clear that you could not trust the layers at all. The assumption that they were all in the right order was wrong. The layers were all jumbled up. The most spectacular example of this is what Maarten called the "pylon nappe"—pylon because it was found near the pylon service road. The "nappe" part requires more explanation.

To a geologist, a nappe is a large slab of rock that has slid away from its original position, a bit like a landslide, but in a peculiar way. As the slab becomes detached, it rolls over on itself like the caterpillar track of a bulldozer, ending up as two layers on top of each other, but with one of the layers completely upside down. This can happen close to the surface, or on the sea floor, or much deeper in the crust involving the movement of very large slabs of rock, several kilometers thick. The word nappe is derived from the French for "napkin," which, when folded in half, looks rather like the twisted layers of rock. One could clearly see this folded "napkin" by tracing out the layers of chert sticking out of the landscape. We spent many hours staring at them with binoculars, marveling at the way the layers snaked right back on themselves, so what looked like two different layers were in fact exactly the same. The importance of this becomes clear when one considers the scale of this repetition, extending for kilometers. Perhaps most of the greenstone belt had experienced the same type of repetition? This would explain why such an apparently thick pile of volcanic and sedimentary rocks could show so little sign of being deeply buried.

My week with Maarten and Isabelle had shown me a new way to read these rocks. I realized that the real story in the rocks might be very different from that indicated by the "standard model." I returned to Swaziland on my motorbike, brimming with excitement.

BACK TO MALOLOTJA

I could hardly wait to test Maarten's ideas with my own observations. As soon as I could, I headed off on my trail motorbike into the heart of the Malolotja Nature Reserve. Now I felt secure and confident—quite different from those feelings of fear and helplessness I experienced on the rutted road to the old barite mine in the camper van. The power of the motorbike was incredible. In the lowest gear, it seemed like it could climb up a vertical wall. And because it only had two wheels, it could safely go just about anywhere—the worst that could happen was that I simply fell over. I decided

FIGURE 2.11 A test of my riding skills. After following a steep track, I reached a remote part of my field area on my brand-new Honda XL185S. (Photo credit: Simon Lamb)

to start by visiting an abandoned gold mine, marked on the map as the She Mine—possibly named after the 1887 H. Rider Haggard adventure novel *She: A History of Adventure* (remember "She who must be obeyed"?). I was following a disused rough track—so overgrown that at times it felt as if I were simply ploughing through chest-high grass. I had chosen to go here because it gave me easy access to the central part of the Malolotja Nature Reserve. According to the available geological map, once I entered the reserve, the track traversed the volcanic rocks of the Onverwacht Group, crossing occasional layers of cherts, all the way to the mine.

All that was left of the mine was a gaping tunnel entrance, rusting equipment, and some large spoil heaps overgrown with bushes. The tunnel lay at the base of a prominent ridge of white sandstone made up almost entirely of tiny quartz grains—a quartzite, to geologists. The quartzite formed a wall of rock, and it looked as though the miners had been searching for gold along the junction between the quartzite and the volcanic rocks, driving a horizontal tunnel to reach it. North of the mine, the landscape opened up. Here, the quartzite formed small rocky hills in a sea of grass.

The overall configuration of the strata was clear to me. At the mine, the strata seemed turned on end, showing up as vertical layers. But in the more open grassland, the layers defined large slabs of rock that sloped into the ground at an angle of about thirty degrees. Looking at these slabs, I thought I could join them all up in my mind's eye as a gigantic trough-like corrugation in the landscape, about five kilometers across and with one side forming a wall made up of vertical layers of quartzite. On the other side of this trough, the layers seemed to curve over again. This way, the layers of quartzite were warped rather like a scaled-up sheet of corrugated iron with its troughs and crests. According to the map, the quartzites were part of the Moodies Group—and from what I could see, they did indeed look like the sandstones from this group that I had seen with Maarten. In other words, they were the top part of the rock pile in the standard model of the Barberton Greenstone Belt.

But there was a real puzzle here: the large mass of volcanic rocks from the Onverwacht Group appeared to be lying *inside* the trough of quartzites and draped *over* the adjacent crest. Yet according to the standard model, these rocks were supposed to be at the bottom of the pile,

Geological map of Malolotja Nature Reserve and adjacent regions, Eswatini and South Africa

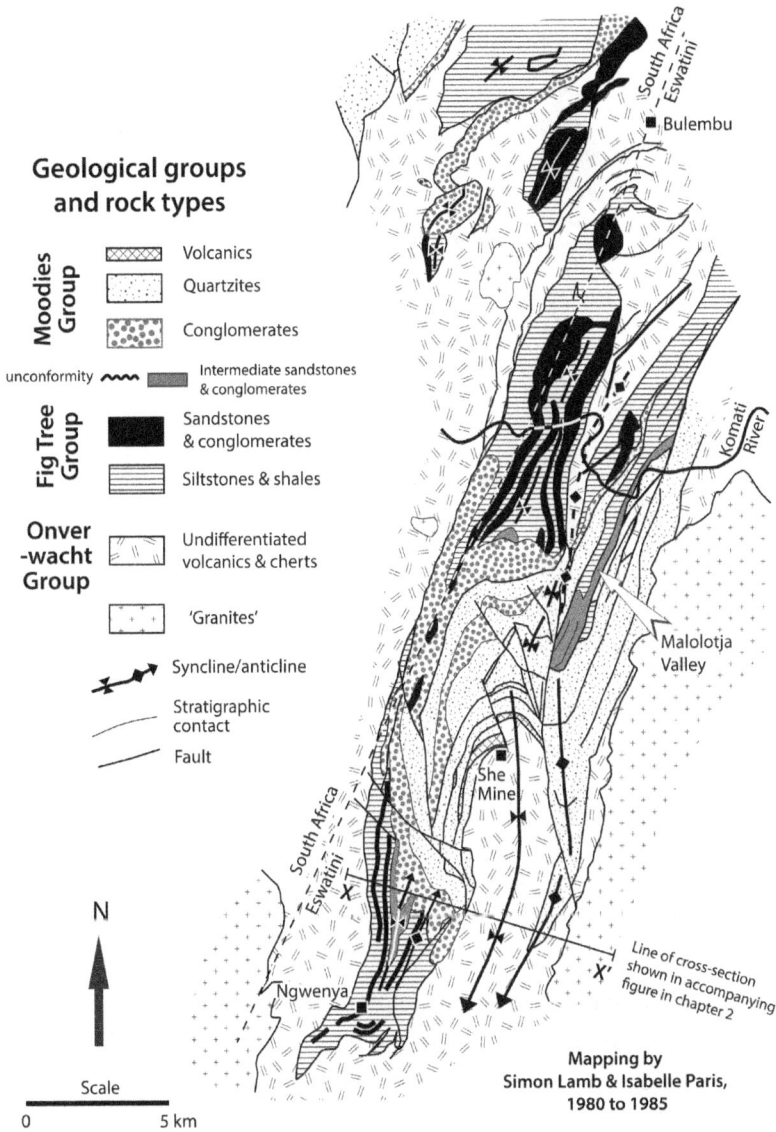

Geological groups and rock types

Moodies Group
- Volcanics
- Quartzites
- Conglomerates

unconformity ~~~~ Intermediate sandstones & conglomerates

Fig Tree Group
- Sandstones & conglomerates
- Siltstones & shales

Onver-wacht Group
- Undifferentiated volcanics & cherts

- 'Granites'

- Syncline/anticline
- Stratigraphic contact
- Fault

Bulembu

South Africa
Eswatini

Komati River

Malolotja Valley

She Mine

South Africa
Eswatini

Ngwenya

Line of cross-section shown in accompanying figure in chapter 2

N

Scale
0   5 km

Mapping by
Simon Lamb & Isabelle Paris,
1980 to 1985

FIGURE 2.12 It's all in a map. The result of nearly 5 years of work by Isabelle Paris and me is this geological map of the Malolotja Nature Reserve in Eswatini and the adjacent region in South Africa, on the southeastern edge of the Makhonjwa Mountains. The sequence of sedimentary and volcanic rocks is divided into three geological groups, and the layers are now folded and sliced up by faults, shuffling the original sequence: see the next figure for a sideways view (geological cross-section) through the southern part of the map. Now that the hard work is done, the map serves as a geological window into the surface of early Earth, about 3.2 to 3.5 billion years ago. (Illustration credit: Simon Lamb)

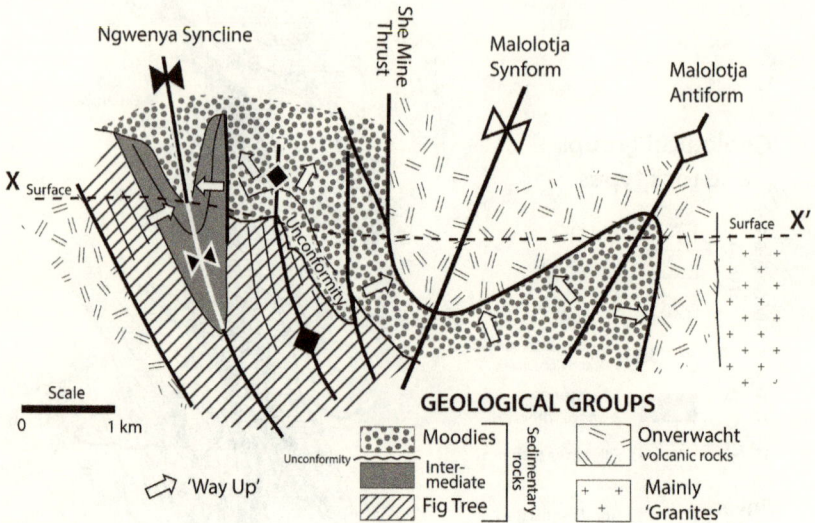

FIGURE 2.13 A sideways view of the bedrock in the Malolotja Nature Reserve. Geological sections like this help geologists visualise the three-dimensional structure of rock layers. Here, the significance of the folds at Ngwenya and the major fault I discovered at the She Mine becomes clear. These features are part of profound Earth movements that uplifted volcanic rocks of the Onverwacht Group and thrust them over much younger sandstones and conglomerates of the Moodies Group, involving horizontal displacements of many kilometres. Later movements compressed these earlier features, so that the faults themselves are folded around the Malolotja synform and antiform. (See glossary for definitions of syncline, anticline, synform, and antiform.) (Illustration credit: Simon Lamb)

certainly not sitting on top of the quartzites. Getting them into their present position would require a huge displacement along a gently inclined fault, known as a thrust fault to geologists, enough to lift the bottom of the pile to the top. In fact, the displacement would need to be at least tens of kilometers—perhaps even hundreds—comparable to those seen today in mountain belts like the Himalayas, where continents have collided. And all this would have occurred before the layers of quartzites were warped into their present orientations. As far as I was aware, nobody else had come up with anything like this before for these rocks. Perhaps I had somehow got it all wrong. I certainly needed more proof. Here was a question I could get my teeth into.

I soon realized that there was a relatively straightforward way to determine if my interpretation was right, and that was to work out which way was up for the slabs of quartzites. In other words, in which direction did the layers get younger? Did they always get younger *away* from the Onverwacht Group, as required by the standard model, or did they get younger everywhere *toward* these volcanic rocks, as I was predicting?

I MAKE A DISCOVERY

The task I set myself was to find those ripples or dunes inside the layers of quartzites, picked out as cross-bedding by subtle changes in the size or color of the sand grains. If I could see some good examples, then they would solve my problem. Evidence for ancient movements in Earth's crust on a truly gigantic scale—perhaps for continental collision and hence early plate tectonics—hinged on the shape of a few dunes or ripples. Such is the power of field geology.

The prospects, however, were not good. It was clear from everything I had read and been told, or seen on existing geological maps, that nobody else had managed to do this. And when I started looking at the quartzites, I could see why. Although I could make out the general layers, forming slabs of rock, the layers seemed internally featureless. I went from one rocky outcrop to another, looking for subtle changes in how the rock surface had weathered. Or I chipped pieces of rock off with my hammer and peered inside. Nothing. As a source of further frustration, large herds of wildebeest grazed this grassland. I remember being trapped on one of these rocky hills by wildebeest. Every time I ventured onto the flat intervening land, the wildebeest bunched together and charged, lowering their heads so that their horns pointed straight at me. At the last minute, they would suddenly wheel around in unison and head off in a different direction. But the overall effect was so intimidating that I did not have the courage to face them, and I had to wait an hour or so until they had decided to move elsewhere.

Here I should tell you something remarkable about field geology: a process called "getting your eye in." When you first start looking at rocks in a new region, you are often confused by distracting detail. It's a strange process—hard to describe—but if you keep looking, the brain gradually starts to apply a filter, so that you only focus on the geological world. It may be that all that is required is a slight change in the angle of the sun, throwing up shadows that reveal what you are looking for. I had just enough experience to realize this, and so I returned the next day and persisted with my search. There was a particular outcrop that caught my eye, but I still could not make sense of it. I passed it several times, and then eventually had lunch sitting on it. As I chewed my sandwich, I suddenly saw what I was looking for: the telltale sign of cross-bedding, picked out by small changes in the size of the sand grains, cross-cutting other similar laminations. It was as if I had taken off a blindfold. I now saw evidence for ripples and dunes everywhere. How could anyone ever have missed them?

FIGURE 2.14 My moment of triumph. I discovered, for the first time, cross-bedding in the Moodies quartzites preserved in the Malolotja Nature Reserve. The critical truncation features, shown by the white arrow, provided the evidence I needed to determine the "way up" of these sandstones (in the direction of the white arrow). (Photo credit: Simon Lamb)

There was no doubt. Wherever I looked, the sandstone layers indeed became younger upwards and *toward* the overlying Onverwacht volcanic rocks, quite the opposite of that predicted in the standard model. The volcanic rocks really had been shoved on top of the quartzites along a huge fault, just like the faults in much younger mountain ranges. I remember shouting aloud to anyone who might hear me—herds of frisky wildebeest, an occasional warthog, and unwelcoming families of baboons—though I don't suppose that any of them would have cared much about my discovery. I had unexpectedly stumbled upon a clue to the existence of very ancient plate tectonics. I will return to this fault in chapter 5. Reading the rocks in this part of the Barberton Greenstone Belt was clearly going to be a painstaking process. But to me it now seemed worth the effort. In the rest of this book I will take you on a series of geological journeys, uncovering the deep history of Earth and linking this to the planet we know today. In the next chapter, I will begin with those volcanic rocks that comprise the "unexpected" Onverwacht Group. But first, here are a few more short stories about my time working in southern Africa.

- § -

MOTORBIKES AND FENCES

After my car accident on the ill-fated visit to the barite mine, it became very clear that if I were going to continue with my project, I would need a safer form of transport to reach the rocks I wanted to study. I noticed a store in Mbabane that sold building equipment and agricultural implements. In the window, proudly on display, was a shiny blue trail motorbike, with huge front forks and rugged-looking knobby tires. For anyone interested in motorbikes, it was a Honda XL185S. Looking at the price, I realized that with the money I had received as a university prize, I could afford it. It turned out that the store had only received delivery of the motorbike that morning, and it was the first one of its kind in Swaziland.

At this point a problem presented itself, which at first seemed insurmountable. It was against the law for the store to sell me the motorbike unless I was a permanent resident of Swaziland, because I would not be able to get the bike licensed otherwise. When I told the store owner I was only there for a short visit, he shook his head. Although I did not realize this at the time, I had several things going for me. First, I was very young and inexperienced. Second, I was desperate. And third, I was not going to give up easily. The store owner agreed to hold the bike for me, if I could get the required documentation from the police by the end of the day. As I left the store, I was pretty sure he thought it was hopeless. I looked at my watch. It was late morning. I immediately headed for the police headquarters in Mbabane.

Police stations tend to attract people of all sorts, and the main one in Mbabane was no exception. Everywhere I looked, there were people shouting and jostling, squeezing into the main hall and arguing with various officials seated behind desks. The job of these officials is to stamp pieces of paper—or at least to let the stamp hover just above the paper, waiting, I suppose, to be convinced that the applicant is entitled to it. I noticed an office with an open door in the far corner. Inside, a policeman in the dark blue uniform of the Royal Swazi Police was working quietly on his own. I forced my way through the crowd, and then stood near the entrance to this office, wondering what to do next. To my surprise, the policeman looked up and asked if he could help me. This seemed too good an opportunity to miss.

My inexperience came to my aid. I passionately launched into my predicament, how I was a geologist working with the Swaziland Geological Survey, and how I had permission from the Swaziland National Trust Commission to study the rocks in the Malolotja Nature Reserve, but I needed a motorbike. As I was talking, he kept interrupting with the question 'What are you?' At first, I didn't realize the significance of this. And then it dawned on me that he was trying to find some alternative way of filling in the daunting space in the motorbike registration form that required my permanent residence number. Finally, he said "Surely, you are a civil servant?" Grasping at straws, I agreed. He took the form and wrote "Civil Servant" in place of the residence number, then found the appropriate stamp and proceeded to authorize his amendment with a thump and a squiggle for his signature. In that brief moment of empathy, ingenuity, and action, he had resolved all my problems.

The store owner had great difficulty believing that I had achieved the impossible. I still wonder today about what happened. I can only think I had encountered an example of straightforward pragmatism, something that I witnessed many more times in Swaziland. Perhaps this was how the king had kept Swaziland safe all those years? Anyhow, the motorbike gave me immediate access to the Malolotja Nature Reserve so that I could explore it on my own. In later years, I purchased an ancient Land Rover to continue this work, and would often camp out for several weeks at a time.

My entire field area was surrounded by a three-meter-high animal-proof steel fence that extended for tens of kilometers, funded by the United Nations Educational, Scientific, and Cultural Organization or UNESCO for short. Under the orders of the king, all people living within the fenced zone had to be moved out. Fortunately, there were only a few remote farms, and the families were relocated to land just outside the perimeter. This relocation was still going on when I first visited the country. It involved numerous journeys along extremely rough tracks by vehicles piled high with roof thatch, wood, and all the possessions of the displaced families. Meanwhile, the fence continued to snake its way through the rugged hills, steadily extended by fencing crews. The man in charge of this operation—also the warden of the reserve—had many years of experience running game reserves in Africa.

I had firsthand experience with the zealousness of the fencing crews a year later, when I was showing some fellow geologists some highlights of my geological discoveries. We were going to look at the southern end of the Malolotja Nature Reserve, the last section to be fenced in, and we started early in the morning when there was still a mist on the hills. Forming a convoy of two Land Rovers, we followed a track that led into the reserve along a narrow high ridge called Ngwenya by the locals, meaning "crocodile's back." After a full day in the field, we were making our way back along the track when we were brought to a sudden halt around a corner by the giant steel fence running right across the road. The fencers were just completing their work, hammering the mesh to thick wooden posts using heavy-duty U-shaped fencing nails. We stared at each other through the fence, as I desperately tried to think of a solution to this awkward problem.

Not far away was the warden, sitting on the hood of his Land Rover and quietly smoking a pipe. He smiled at me and shrugged his shoulders. Then he told the fencers to pull down the fence so that we could get through. It was an embarrassing moment, but the warden agreed it was rash of him not to let me know, and I apologized profusely for causing him so much trouble. My companion, Maarten de Wit, broke the tension with a loud laugh and we parted saying that we should talk more about it over a beer sometime. And the warden did finally give me unlimited access to the reserve, handing over a large bunch of keys to its many padlocks.

Getting in and out of the reserve was always an ordeal, as the fence followed a line on a map that did not always follow logical topographical features. Accessing some parts of the reserve involved driving along a road that crossed the fence line several times. Each time, I would have to clamber out of my vehicle, search through the bundle of keys for the one that would unlock that particular padlock, then wrestle with a long chain wound again and again around the gate. It was always a relief to pass through the last of the gates so that I could start looking at the rocks poking out of the landscape and begin "geologizing" in earnest.

Act II

Act II

3 | World of Oceans

Deep diving: exploring the bottom of the earliest oceans

In the late 1960s, two brothers were carrying out geological field work on the southern flanks of the Makhonjwa Mountains, about thirty kilometers south, as the crow flies, of the old mining town of Barberton. The brothers, Morris and Richard Viljoen, were difficult to tell apart, being twins. They were working on their postgraduate degrees at the University of the Witwatersrand in Johannesburg. Their task was to produce a geological map, sorting out the different types of volcanic rock in the region. In July 1969, the results of this work were presented to the world at a major international conference in Pretoria, the capital of South Africa. This took the geological world by storm, and the names "Viljoen and Viljoen" are enshrined in the scientific literature, with the region where they worked now a UNESCO (United Nations Educational, Scientific, and Cultural Organization) World Heritage Site. I met Richard Viljoen about eleven years later, this time in the newly established Zimbabwe, and he was still trying to make sense of what he had found with his brother. In this chapter, I will describe how their discoveries have opened a window onto early Earth, revealing the nature of the earliest oceans and the way they were created. All this has made the Komati River famous to geologists throughout the world, because much of the Viljoen twins' work was carried out along its banks.

ON THE BANKS OF THE KOMATI RIVER

Komati (or *Nkomati*) is the Siswati word for a cow, alluding to the river's role as a lifegiving source of water in the same way that a cow gives milk to nurture its young. In the dry season, during the Southern Hemisphere winter, you can easily wade across it. But during the wet season, in spring, it can be a raging torrent overflowing its banks. The river flows off the Highveld of South Africa, following a convoluted course that skirts the southern

FIGURE 3.1 The Komati River gorge cuts nearly 1,000 meters (3,000 feet) down through the rocks of the Barberton Greenstone Belt in the Malolotja Nature Reserve, Eswatini. (Photo credit: Simon Lamb)

part of the Makhonjwa Mountains, before entering a narrow gorge at the border between South Africa and Eswatini. Here, in the Malolotja Nature Reserve where my project was based, the river has cut down nearly 1,000 meters through the mountains, allowing it to continue its long journey to the Indian Ocean.

Upstream of the gorge, the landscape is more open, and the Komati River and its tributaries flow over relatively "soft" volcanic rocks. This is a dry landscape of long grass, small rocky hills called kopjes, and stands of thorny scrub and protea, the national plant of South Africa. The land had been divided up by Boer settlers into farms. Geologists have taken the name of one of these, the Onverwacht farm (after the Afrikaans word for "unexpected"), as a convenient label for the underlying volcanic rocks, assigning them to the Onverwacht Group. This, you may recall, forms the bottom part of the thick pile of volcanic and sedimentary rocks in the standard model of the Barberton Greenstone Belt. For the sake of simplicity, I will continue to refer to the volcanic rocks in this way, although it is now recognized that they have been profoundly sliced up by movements along faults. Over time, the Komati River has worn away the Onverwacht Group, creating wide rocky pavements on its banks. Intricate details of the rocks are beautifully exposed here, providing the Viljoen twins with exceptional opportunities to study the volcanic activity of early Earth.

The rocks were dark and heavy, and they sometimes had all the features one would expect for lava that erupts underwater. We now know in great detail what happens during these eruptions because this type of eruption has been witnessed by scuba divers—although it is very dangerous to get close because the lava is at a temperature well over 1,000 degrees Celsius, and the adjacent superheated water is capable of causing severe burns. When molten lava emerges from an underwater crack, it cascades down any slope in the seabed, rapidly cooling on the outside to form a solid crust. But this crust is continually broken by the pressure of the still-molten lava inside, and new buds of hot lava burst out, and so the process repeats itself. This builds up into a gigantic pile of rocky blobs, or "pillows" of lava. For this reason, they are called pillow lavas. It is a bit like squeezing toothpaste out of a tube that keeps getting blocked, producing a jumble of toothpaste lumps.

FIGURE 3.2 Maarten de Wit examines exceptionally well-preserved examples of pillow lavas (komatiitic basalt) from the Onverwacht Group on the banks of the Komati River. These show all the features seen in modern underwater eruptions on the deep seafloor. (Photo credit: Simon Lamb)

The Viljoen twins found beautiful examples of these pillows, nestled together just as they are in modern underwater eruptions. Occasional layers of fine sediment indicated periods of volcanic quiescence, when mud had time to settle out. There were also abundant lava flows that were not pillows, but instead formed massive layers that could be traced through the landscape. Despite their great age, there was nothing particularly unusual about any of this for a thick pile of volcanic rocks. But when the Viljoen twins analyzed them back in the laboratory, it became clear that they had found a type of rock new to science. In fact, the existence of this rock sparked an intense scientific debate about whether early Earth was a profoundly different place compared to the planet we know today. To understand this debate, and why it is so important for working out the conditions on early Earth, one needs to know what geologists were expecting based on what they had learned about the origins of volcanic rocks.

A ROCK FOUNDRY

I described in chapter 1 how a volcanic rock called basalt is continually erupting at the mid-ocean ridges that run down the middle of the world's oceans, building up an oceanic crust that covers over two-thirds of the planet's surface. This makes basalt the most common volcanic rock, and possibly the world's most common rock altogether. Occasionally, beautiful greenish lumps of mantle rock, full of crystals of olivine, are caught up in the lava. The presence of these lumps shows that basalt must ultimately be derived from the underlying mantle. Thus, given the abundance of basalt, the story of volcanic rocks starts with what happens in the mantle. This has led to a major research effort. Some geologists have scrutinized the solidified remains of ancient bodies of molten rock—magma—while others have created magma by melting rock in the laboratory or by investigating molten lava in active volcanoes.

It is clear that volcanic rock begins its life as magma. Thus, basalt must be derived from molten mantle rocks, and so working out the way mantle rocks melt is the key to understanding how basalt is created. Mantle rocks, however, are not single substances, but rather a mixture of different minerals that are bonded together. Therefore, unlike pure metals, they do not melt all at once, but in stages according to when different minerals reach their particular melting points. In the laboratory, a typical mantle rock such as peridotite begins to melt in an open crucible at around 1,100°C—similar to the melting point of gold—but it does not become fully molten until it reaches a higher temperature, around 1,750°C, not far from the melting point of platinum. Under the huge pressures in the mantle itself these temperatures will be much higher. For this reason, despite the fact that the interior of Earth is thousands of degrees hotter than the surface, there are only very limited parts of the mantle that have the right temperature and pressure conditions for melting; the rest is solid rock. I will deal with the question of what happens in the mantle to make the right conditions further on in the chapter. However, even here, only a small fraction of the rock will have melted.

This brings us to another important idea: the composition of the molten portion of a rock—the magma—will depend on how much of the rock has melted. Thus, if the mantle is only partially molten, its composition will not be the same as that of the solid mantle. This goes back to the fact that rocks in the mantle are mixtures of different minerals, each with its own individual melting behavior. The significance of this is clearer if one knows how geologists have traditionally gone about classifying igneous rocks—volcanic rocks, of course, are just igneous rocks that have erupted from a volcano as lava or ash, whereas the rest is stuck in the crust as bodies of solidified magma. It turns out that the compositions of igneous rocks have something in common: they are determined by minerals known as silicates, which contain a high proportion of silicon and oxygen and contribute silica (that is, silicon dioxide) to the rock composition. Geologists, therefore, have classified igneous rocks according to their silica content by weight.

These divisions are significant for understanding the origins of volcanic rocks, although I will avoid the technical names in the following discussion. To start with, in large quantities, silica makes magma sticky. Thus, a silica-rich lava is stickier than a silica-poor one, and, for this reason, will tend to erupt more explosively. But silica also plays a role in the way rocks melt and crystallize. Thus, melting a small fraction of a mantle rock produces a magma that contains more silica than the mantle rock itself. This is what goes on beneath the mid-ocean ridges today, where magma with the composition of basalt—containing about 50 percent silica—is effectively sweated out of mantle rocks that contain only about 43 percent silica.

It is not only the amount of silica that changes during melting. The amounts of other elements do as well. For convenience, these elements are also quantified in terms of the proportions by weight of their oxides. The amount of magnesium oxide is particularly interesting in relation to mantle melting because, unlike silica, magma usually contains considerably *less* of it than the mantle does. In other words, it requires large amounts of melting to boost this level in the magma, and so the amount of magnesium oxide is a sensitive measure of how much of the mantle has melted. This has led to another way of categorizing igneous rocks, based on how much magnesium oxide (and iron oxide) they contain. A typical basalt contains moderate amounts of magnesium oxide—around 10 percent by weight.

To a geologist, this makes a basalt "mafic"—a term derived from 'ma' for magnesium, and 'f' for ferrum, the Latin word for iron. However, mantle rocks are "ultramafic" because they are very rich in magnesium oxide (up to 50 percent by weight). So, we have a twofold difference here between the mantle and the basalt it produces when it melts: an increase in the amount of silica on one hand, and a decrease in the amount of magnesium oxide on the other.

The world is never simple, and there is more to turning magma into volcanic rock than one might expect. First, the magma has to escape from the mantle. Here, the weight of the rocks plays an important role because it forces the magma out, just like squeezing water from a sponge. This way, the magma can begin its journey toward the surface. Beneath the overlying volcanoes, magma is channeled through a complex plumbing system. This plumbing system can get blocked, causing magma to build up at depth and then be released suddenly—and often violently—during a volcanic eruption. Thus, the lava that spews out of a volcano is not always coming directly from the mantle, but from a reservoir or pool of magma slightly higher up. Geologists call this reservoir a magma chamber, and here the magma may sit around for long periods—perhaps years to hundreds of years, or even longer. During this time, the magma slowly cools and begins to crystallize. And it is this crystallization, by selectively leaving or taking particular components, that can refine the remaining magma into the compositions of a wide variety of other volcanic rocks.

So far, so good. The understanding of melting and cooling of rocks that I have just described has served geologists well, helping them explain the origins of almost all lavas erupting today at Earth's surface. The central idea is that melting of the mantle produces volcanic rocks—mainly basalt—that are very different from the mantle itself.

SOMETHING UNEXPECTED

We are ready to return to the rocks discovered by the Viljoen twins in the late 1960s. I have already described how they were examining the volcanic rocks along the banks of the Komati River, beautifully exposed in the worn river pavements. They had thought at first that these were typical lavas of basalt, some of which had erupted underwater, much as one would see in a volcanically active region like Iceland. But, on closer examination, many of the lavas turned out not to be basalts at all, but completely unexpected lavas that were far more like the rocks in the mantle than the crust. I will soon come to their significance, and why they caused such a stir in the geological community. But first let me describe their unusual features.

The lavas were sometimes full of the mineral olivine, which is the main component of the mantle, just below the crust. Silica concentrations were unusually low, similar to those of mantle rocks. And importantly, many of the lavas were ultramafic, with up to 30 percent magnesium oxide by weight, or as much as *three* times what would be expected in a typical basalt. As I have mentioned, such extreme amounts of magnesium oxide are normally only found in the mantle. It looked to the Viljoen twins as if the mantle had somehow erupted as a lava on Earth's surface. The Viljoens called their new rock a komatiite (pronounced: koh-MAH-tee-ite), the rock of the Komati River, and the name has now been adopted by geologists for any volcanic rock with the same key features—thus, to be a komatiite, the rock has to have crystallized from a magma that contained *at least* 18 percent magnesium oxide by weight. The twins also found less 'extreme' examples, closer to basalt in composition, which they called komatiitic basalts. There were some normal basalts as well. The "unexpected" Onverwacht Group had certainly lived up to its name!

The chemical composition of the komatiites also made sense of something else that had puzzled the Viljoens. When molten, these lavas would be predicted to be less sticky than basalt (and komatiitic basalt), and so capable of flowing faster and farther, covering a wide swath of the ancient landscape or seafloor with a thin veneer of "mantle-type" rock. This explained why some individual lava flows were very thin and laterally extensive. In addition, the komatiites were slabs of lava, whereas the stickier komatiitic basalts and true basalts tended to have lumpy shapes, forming the pillow lavas. Even more striking was the unique shape of the individual crystals of olivine in the komatiites. Olivine typically forms crystals with shapes and sizes similar to coarse sand or even fine gravel.

FIGURE 3.3 A komatiite volcanic rock from the Onverwacht Group with characteristic long, needle-shaped crystals of olivine (above the pencil), forming the so-called spinifex texture— named for its resemblance to spinifex grass in Western Australia. (Photo credit: Simon Lamb)

But the olivine crystals in the komatiites (and komatiitic basalts) could be long, needle-like structures, reaching tens of centimeters in length and fitting together as intermeshed sheaves. This was later called a spinifex texture because the crystals resembled the spiky spinifex grass that grows in Western Australia, where komatiites were soon found in similarly ancient rocks. And it was further evidence that there was something strange about these rocks.

It was not until many years after I had started working in the Barberton Greenstone Belt that I got the opportunity to see for myself the actual outcrops of rock in South Africa that the Viljoen twins had brought to the attention of the geological world. I was returning to Barberton with a BBC film crew to make the Earth Story documentary series about the evolution of Earth. Our visit to this region was part of a journey back to the beginning, much as I am trying to do in this book. It was Maarten de Wit who took me to some of the famous exposures of komatiite, complete with spinifex textures, as well as to the komatiitic basalts with their pillow shapes. I could not have had a better guide. After our first meeting many years earlier, at the start of my own doctoral research in the Makhonjwa Mountains, Maarten had continued exploring the geology throughout the Barberton Greenstone Belt—first while based at the University of the Witwatersrand in Johannesburg, and then at the University of Cape Town. And the focus of his research was the Onverwacht Group. The degree of preservation in the rocks Maarten showed me was truly staggering—I have a photograph of him sitting on pillow lavas that look as though they have just recently erupted on the seafloor. It is hard to believe that these are relics of early Earth's surface, only a quarter of the way through our planet's long history.

Since then, over the eons of geological time, the tectonic plates have had time to wander all over the world, mountain ranges and oceans have come and gone, life has evolved, and the landscape of this part of southern Africa has been worn away by the forces of erosion. Paradoxically, perhaps, this erosion is one reason why so much of the greenstone belt is so well preserved—the rocks here have been exposed only relatively recently (from a geological point of view) to the inevitable weathering that takes place close to Earth's surface.

Typical sequence of lavas in Komati Formation of the Onverwacht Group

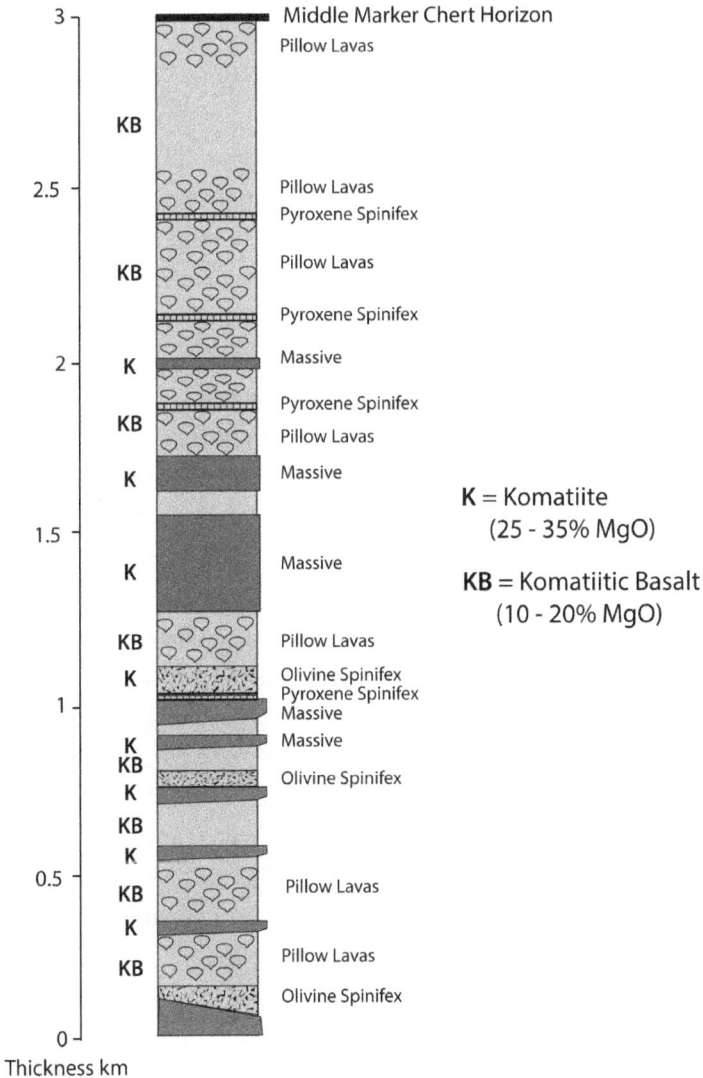

FIGURE 3.4 A geological log, based on geologist Jesse Dann's detailed work, illustrating the volcanic rock sequence exposed along the banks of the Komati River. These rocks were first brought to the attention of the world by the Viljoen twins in the late 1960s, called by them the Komati Formation, which is part of the Onverwacht Group. The rock sequence consists almost entirely of layers of pillow lavas and massive flows, known as komatiites and komatiitic basalts, characterized by an unusual chemical composition: extremely high in magnesium oxide and relatively low in silica. Clusters of needle-shaped olivine or pyroxene give some of the lavas the distinctive spinifex texture. (Illustration credit: Simon Lamb)

AN IMPOSSIBLE ROCK

You may still be wondering what all the fuss was about. What does it matter that the composition of the komatiite lavas was so similar to that of the mantle? This similarity was in fact the crux of the problem, because it seemed to contradict all the hard-won understanding I described earlier about how volcanic rocks form. Given the conditions of Earth today, the fundamental physics and chemistry of melting and crystallizing rocks predict that lavas should be significantly different from the underlying mantle, from which they are ultimately derived. It seems we have an impossible rock. But we need to be careful here. We have a rock that may *now* be impossible, but clearly not when the komatiites in the Barberton Greenstone Belt erupted over three billion years ago. So, what has changed? Not the physics and chemistry of melting—it must be the nature of Earth's mantle.

Geologists tackled the problem of komatiites head-on, trying to produce them through melting typical mantle rocks. It was clear that the eruption temperature of a komatiitic magma (1,450°C or higher) is significantly hotter than that of a basaltic magma (around 1,200°C). If you had witnessed an eruption of komatiite, you would have seen a bright white liquid, so hot that you would not have been able to get anywhere near it, and so turbulent it would have splashed its banks with droplets of magma like a fast-running mountain stream. Molten basalt is only red hot, and in comparison to liquid komatiite, flows much more slowly. And producing a komatiitic lava would require melting a much larger proportion of mantle rock. The genesis of a typical basalt involves melting about 10 percent, on average, of the underlying mantle. For a komatiite, it would have to be more—perhaps as much as 50 percent. This way, the molten portion would be closer in composition to the rock being melted. And there could be no prolonged cooling and crystallization in a magma chamber—the magma would need to erupt not long after it left the mantle.

However, not all of this magma would reach the surface, and what was left behind would undergo a longer history of cooling and crystallization, just as I described when talking about the rise of magma in the complex plumbing system beneath a volcano. Here is a way to change the composition of the remaining magma, and it might be how the lavas of komatiitic basalt and normal basalt were produced. Thus, the different volcanic rocks found by the Viljoen twins in the Onverwacht Group might all be ultimately linked to komatiite, further demonstrating the importance of this type of rock in the volcanic activity of early Earth.

Geologists soon concluded from all this that, when komatiites erupted, the mantle was melting much deeper down because Earth's interior was hotter back then—to be more precise, several hundred degrees Celsius hotter on average. And so the origin of komatiites became a question of mantle temperatures during the planet's early history. At first, there seemed to be a simple answer, since it is widely agreed that when Earth first formed—over 4.5 billion years ago—it was largely molten, or at least, significantly hotter, heated by the impact of cosmic dust as it collided and compressed to form the planet's interior. However, studies of Earth's subsequent cooling from this hot birth soon showed that a billion years into the planet's history—around the time the volcanic rocks studied by the Viljoen twins erupted—the mantle, although still hotter than today, was not hot enough to produce komatiites. in fact, it was up to 200 degrees Celsius too cool. Conditions in the mantle were clearly closer to those needed to produce komatiites, but something else must have been going on.

While geologists were grappling with this problem, another revolution was taking place in our understanding of Earth's deep interior. Material scientists had long known that rocks at high temperatures and pressures, despite being solid, are still capable of flowing. Geophysicists were now investigating what this flow of rocks might look like inside the mantle, and their results were giving us new ways of thinking about the origin of komatiites. Let me explain.

A GIGANTIC LAMP

Remember those colorful lava lamps that were so popular in the 1960s and '70s? You can still buy one today—a glass container filled with a mixture of colored water and modified paraffin wax. It works because there is a hot electric bulb at the bottom of the container. The paraffin wax near the bulb heats up and expands, becoming less dense and so it floats upward

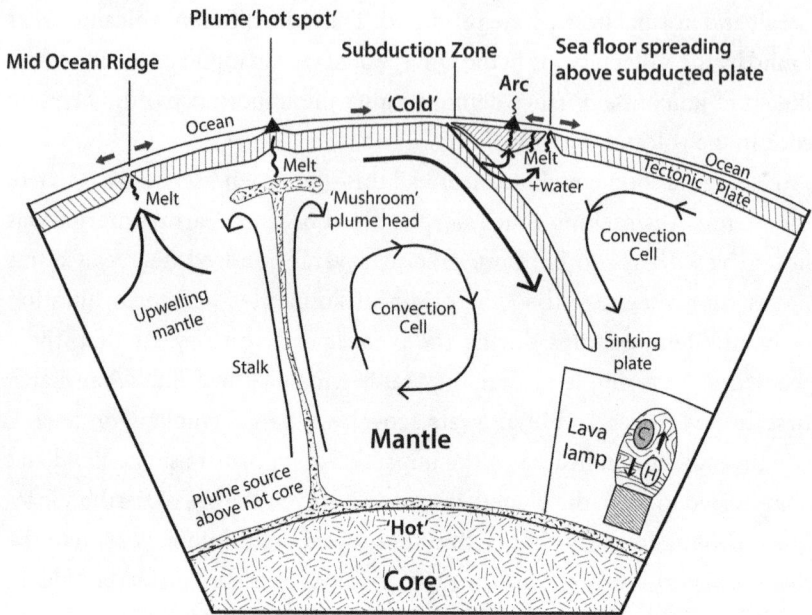

FIGURE 3.5 The workings of Earth's deep interior. Rocks at the base of the mantle are heated by the hot core but cool nearer the surface, creating a rising and sinking flow pattern called convection rather like that in a lava lamp. The cold top of the mantle forms tectonic plates, which move apart above upwelling mantle at mid-ocean ridges and sink back into the mantle at subduction zones. Occasionally, thin vertical jets of hot mantle, called plumes, rise from the bottom of the mantle. Localized mantle melting produces basaltic (or in some places, boninitic) magma that feeds overlying volcanoes. This melting occurs in the hot mantle upwelling beneath mid-ocean ridges and other seafloor spreading centers, in rising plume heads, and in mantle above sinking tectonic plates enriched with water carried down in a subduction zone. In the hotter mantle of early Earth, at least some of these melting regions may have produced komatiites and komatiitic basalts. (Illustration credit: Simon Lamb)

through the water. As it moves away from the bulb, it cools and contracts, now becoming more dense, and so it starts sinking again, creating a perpetual motion of rising and falling colorful blobs. This motion is an example of a more general behavior of fluids heated from below, known as convection. Convection takes place every time you heat a saucepan of water on the stove or turn your electric kettle on.

The eminent British physicist Lord Rayleigh—who received the Nobel Prize in 1904 for his discovery of the element argon—showed that a fluid heated from below will undergo convection if it has certain physical properties, which can be combined into a single value known as the Rayleigh number. This is really just a measure of the relative importance between the two ways that thermal energy—heat—can move through a fluid: either carried in moving hot blobs of fluid, as in a lava lamp, although this will be resisted by the fluid's stickiness; or by conduction, in the same way that the solid walls of your house let heat escape from the warm interior to the cold exterior. The minimum Rayleigh number required for convection is about one thousand, a mathematical way of saying that convection occurs when blobs of rising or sinking fluid can move fast enough to dominate the flow of heat.

It turns out that Earth's mantle shares important features of a lava lamp, or indeed, a heated saucepan of water. Instead of an electric bulb or stove, Earth's mantle is heated by the underlying hot core, although it also contains radioactive elements that provide more heat due to their radioactivity. We know that the core is a body about the size of Mars at the center of the planet, made of iron and nickel, with a temperature nearly as hot as the surface of the Sun (see chapter 7 for more). This makes the bottom of the mantle, at a depth of about 2,900 kilometers, thousands of degrees hotter than the top. But the mantle is still solid rock, kept this way by the huge pressures in Earth's interior. Importantly, though, the rocks here can flow—albeit very slowly. This can be directly observed in Scandinavia and Canada, where huge ice sheets once pushed down Earth's surface. The displaced mantle rocks are now flowing back, and the land is slowly rising again. The key question is whether they can flow easily enough to also make convection possible.

Detailed calculations show that not only is the mantle's Rayleigh number high enough for convection, but it is actually far more than is needed—well over a million. The mantle really *wants* to convect; there is nothing reluctant about it! This is a truly remarkable result when one considers that the stickiness of mantle rock—technically defined by what is called viscosity—is about a million billion billion times greater than that of water. And yet the Earth's mantle, saucepans of hot water, and lava lamps are all capable of convection!

RISING JETS

Experiments with heated fluids, along with computer calculations using the physical laws governing fluid flow, show that convection in Earth's mantle can happen in a number of different ways. In fact, plate tectonics is a form of convection, involving upwelling mantle rock beneath the mid-ocean ridges and its sinking in subduction zones. However, experiments and computer calculations show that the flow is more complicated than this: local jets of hot rock regularly detach themselves from near the base of the mantle and rise toward the surface, like the lava lamp's ascending blobs. The speed of the jet is very fast, geologically speaking—it might take only a few million years to make its way through the entire mantle. Near the surface, the jet spreads out sideways creating a mushroom-shaped head that can be thousands of kilometers across. Eventually, it starts to sink back into the mantle at its edges like the hanging lip of the mushroom. Upwelling jets can be imaged today, deep in the mantle, using essentially the same technique as a medical CT scan. This involves carefully analyzing when seismic vibrations from earthquakes are first detected at various locations around the world. The telltale sign of a jet is the later than expected arrival of the seismic vibrations, because the vibrations have been slowed down as they pass through the slightly hotter mantle of the jet. Viewed this way, the mantle contains numerous vertical arms of unusually hot rock radiating out from the top of the core, a bit like the pins sticking out of a pincushion.

The rocks inside one of these rising jets, or plumes as they are usually called, are calculated to be a few hundred degrees hotter than the surrounding mantle. This reflects the higher temperatures at the bottom of the mantle, closer to the core, where the plume ultimately came from—the plume is rising so fast that its interior does not have enough time to cool down to the temperature outside. But deep in the mantle, the weight of the overlying rocks pressing downward creates colossal pressures that are tens to hundreds of thousands times greater than atmospheric pressure, and pressures like these keep the plume almost entirely solid. However, as the plume approaches the surface, the pressure becomes low enough that its hotter interior can begin to melt in significant amounts. The magma rises all the way to the surface, erupting as lava in the overlying volcanoes, such as those in Hawaii, where an underlying plume has been detected.

Plumes provide a way to create localized conditions in the mantle that are suitable for melting, even if the surrounding mantle is not hot enough. Here is a possible mechanism for forming komatiites inside early Earth, because if the mantle at that time was generally hotter, then the even higher temperatures within a plume might have been sufficient to produce komatiitic magma—plumes today are only hot enough to generate basaltic magma. In this case, komatiites are the result of an extraordinary journey of very hot plumes of rock, which have risen through the mantle all the way from the top of the core. As I have already described, these plumes are part of the natural behavior of Earth's mantle, as are plate tectonics. There is no reason to think that plumes would not have existed in the hotter interior of early Earth.

It seems that we have finally solved the problem created by the discovery of those strange volcanic rocks in the Makhonjwa Mountains back in the late 1960s. You might be surprised to learn, then, that there are some geologists who think this is not the solution—or at least, not the only solution. It turns out there is another, quite different idea about the origin of komatiites. An eruption from a strange volcano in a remote corner of the Pacific Ocean points to plate tectonics, not plumes, as the mechanism for making komatiites.

- § -

WEST MATA COMES TO LIFE

Back in 2008, sightings of a large floating mass of volcanic ash were reported by passing ships about 750 kilometers northeast of Fiji, and 250 kilometers southwest of Samoa. This was soon pinned down to an eruption from an isolated underwater volcano called West Mata. The following year, a research vessel with a remotely operated underwater vehicle (ROV) on board was sent to investigate further. The ROV dove to a depth of about 1,000 meters and filmed lava pouring from the volcanic flanks, forming the characteristic pillow shapes that occur when the outer skin of the lava is quenched by ocean water. Near the eruption site, communities of deep-sea shrimp and microbes clustered around vents of hot water. Such was their abundance that one of these vents was named Shrimp City. Samples of the erupting lava were collected by the robotic arms of the ROV and brought back to the surface.

Laboratory analyses soon established that the West Mata volcano was erupting a very unusual type of lava called boninite. The lava is named after the Bonin Islands, farther north and closer to Japan, where it has also been found—this time as part of older volcanic eruptions that occurred about 45 million years ago. Boninite was just beginning to make a name for itself in the geological community. It contains unusually large amounts of magnesium oxide, reaching 20 percent or even slightly more by weight. To geologists, this makes it a close relative of the komatiites—the komatiitic basalts in particular—because one of the defining characteristics of komatiites is their very high magnesium oxide content. Boninite is clearly spewing out of active volcanoes today, although these eruptions are very few and far between.

A clue to the origin of the boninites lies in what is happening, geologically speaking, next door to West Mata volcano. Here, a trench in the seafloor reaches depths of nearly eight thousand meters, where the vast Pacific Plate is sliding westward beneath West Mata volcano in a subduction zone—in fact, this motion is about as fast as it gets for tectonic plates. What happens next to the sinking plate has been worked out by geologists from a wide range of evidence. To begin with, it is transporting water that became trapped in the crust as it lay at the bottom of the ocean, either stored in fractures or bound within minerals in the rock. But the increase in temperature and pressure deep in the subduction zone eventually drives the water out of the rock. The extremely rapid rate at which the seafloor is sinking means that exceptionally large quantities of water are continually being flushed through the subduction zone. The water, being less dense, rises and seeps into the overlying mantle, making it unusually "wet."

Overlying this "wet" mantle is West Mata volcano itself. It rises from seafloor that is geologically very young, created by the continual eruption of lavas at a seafloor spreading center—in other words, at a mid-ocean ridge, but without the connotation of being in the middle of an ocean. The seafloor here is also being further pulled apart along fault lines by this spreading. All this is the surface expression of an upward and outwardly diverging flow of the underlying mantle rocks, rising to fill the space created by seafloor spreading. As these rocks rise, they undergo a reduction in pressure, rather like the decompression experienced by an underwater diver swimming back to the surface. The rocks will eventually reach pressure conditions low enough for melting, given their temperature—this is, in fact, the fundamental reason why there is any volcanism at all where the tectonic plates drift apart at spreading centers. But beneath West Mata, there is also the rise of water that has seeped out of the sinking plate. The presence of this water has a profound effect, triggering additional melting in much the same way that antifreeze or salt causes ice to melt. And it is these exceptional amounts of melting in the mantle that are needed to produce the boninites with their high concentrations of magnesium oxide.

Imagine all this happening on early Earth, when the mantle rocks were hotter to begin with. This might be the furnace in which komatiites were forged, at or near a seafloor spreading center above a subducted plate. An important part of the process is the introduction of water into the mantle, which actually causes it to melt without any change in temperature or pressure. All the experiments on producing komatiites had assumed that there was virtually no water in the mantle. It turns out that the komatiites from the Barberton Greenstone Belt do contain water—up to 3 percent by weight. Recently, the Russian scientist Alexander Sobolev and his coworkers managed to show that some of this water had come from the early oceans. By carefully examining individual olivine crystals in the rock, the team found microscopic blobs of frozen magma containing bubbles of water with the characteristic chemical signature of seawater. To my mind, the simplest explanation for how this water got into the magma is that it was carried down into the mantle with the sinking seafloor in a subduction zone, before being caught up in upwelling mantle that started to melt.

Komatiites seem to show signs of subduction in the rest of their chemical composition. In particular, elements that make up only a very small part of the rock—known as trace elements—can serve as fingerprints, so to speak, of the sequence of events that led to the creation of the magma and the crystallization of the rock. These elements have names that were made famous in the Tom Lehrer song called "The Elements"—you may recall cesium, niobium, lutetium, scandium, tantalum, and thorium, to name a few key ones. Unfortunately, the fingerprints are not always definitive, and there is debate among geologists about whether they can be reliably used in such ancient and potentially altered rocks. Nonetheless, Maarten de Wit and his colleague Harald Furnes at the University of Bergen argue that many of the komatiites in the Barberton Greenstone Belt display all the characteristic fingerprints of subduction.

We have now completed yet another long and tortuous journey in search of the origins of komatiites, this time ending up at a spreading center above a subduction zone. In other words, this is still a world shaped by plate tectonics, similar to parts of the Pacific Ocean today. It goes without saying that there is no consensus among geologists about whether we have arrived at the right destination with this alternative view of komatiites, and the debate rages back and forth. I will argue later on in this chapter that *both* these views of komatiites are likely to be correct, with either plumes or subduction being responsible at different times during the eruption of the volcanic rocks in the Onverwacht Group.

OPHIOLITES

Our quest to understand the strange volcanic rocks in the Onverwacht Group is turning into a much wider investigation of the origin of the crust beneath the oceans of early Earth— sorting out whether it was the product of mantle plumes, seafloor spreading and subduction, or all of these. Thus, even if we accept the existence of plumes and subduction on early Earth, we need much more evidence that seafloor spreading played an important role in the creation of this crust. Looking for evidence for or against seafloor spreading during this time will dominate the next stage of our quest.

Curiously enough, the evidence lies not only in the ancient bedrock of the Makhonjwa Mountains, but also in much younger rocks exposed in the mountains of northern Oman and Cyprus. This might seem rather counter-intuitive, but these younger rocks are the most extensive and well-preserved examples of the ophiolites I mentioned in chapter 1. They have been directly compared with images of the rocks beneath the ocean floor, obtained by geophysicists using a wide variety of techniques to probe the subsurface. Putting all the evidence together, there seems little doubt that the ophiolites in Oman and Cyprus are relics of geologically young oceanic crust that formed around 95 million years ago at seafloor spreading centers, even if geologists have still not completely figured out how they ended up in these mountains. An even younger example is Macquarie Island between Antarctica and New Zealand: a piece of oceanic crust about 10 million years old that has been forced up and tipped on end. These sites provide rare opportunities to directly observe rocks that would otherwise be far beyond our reach, several kilometers beneath the seafloor.

We can use an ophiolite as a model for a typical slice through the oceanic crust today. The top part is sediment, commonly consisting of chert that is

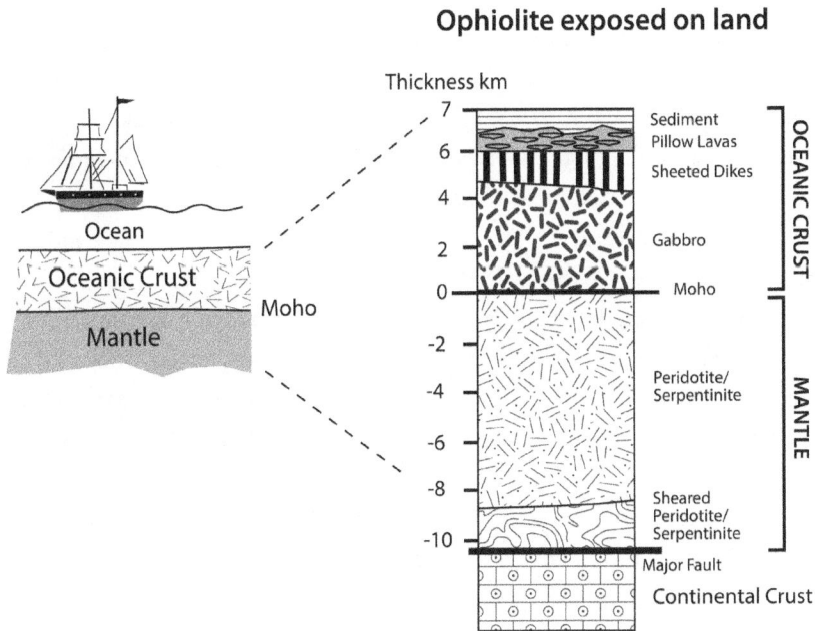

FIGURE 3.6 The crust beneath the oceans can be studied in rare locations on land where it has been pushed up to form ophiolites. These reveal the distinctive sequence of rocks that make up the oceanic crust. (Illustration credit: Simon Lamb)

composed mainly of the remains of microscopic ocean-dwelling organisms called radiolaria. The skeletons of radiolaria are made of silica, and when they die, the skeletal remains sink to the ocean floor. Over time, the remains are gradually buried as more skeletons rain down above them, and they recrystallize to form a cherty or flinty rock. But if you look at this chert under the microscope, you can still see the shapes of the original radio-larian skeletons. Sediment deposited in water shallower than about 4,000 meters may also contain the remains of organisms with calcium carbonate shells or hard parts. But where the ocean is deeper, the high pressure causes calcium carbonate to dissolve in the water column as the dead organisms sink to abyssal depths, leaving only silica skeletons to accumulate on the seafloor. In addition, fine particles of rock or ash may be carried in by dis-tant rivers, or blown in by the wind, eventually settling out as mud and silt. Overall, a layer of sediment up to a few hundred meters thick accumulates in the deep ocean—it is too far away from land for there to be much more.

The sediment rests on basalt, which erupts underwater along spreading centers, forming bulbous, blobby shapes—resembling a stack of pillows—that I have already mentioned. These are the pillow lavas. The pillow lavas build up a layer about half a kilometer to one kilometer thick. They are fed by a basalt magma that rises through vertical cracks. The magma left behind in these cracks also solidifies, forming walls of basalt up to a few meters wide—features that geologists call dikes. The dikes are the conduits that channel the magma toward the surface from deep reservoirs located several kilometers below. It is like a gigantic plumbing system, pumping liquid rock to the surface. Over time, so much magma passes through these conduits that younger dikes can be squeezed between older ones, creating adjacent walls of rock that resemble closely packed sheets of plywood stood on end. Geologists call them sheeted dikes.

At this stage, with these references to both sheets and pillows, you would be forgiven for thinking that geologists have their sleeping arrangements very much on their minds. Tongue in cheek, my colleague Maarten de Wit titled two of his scientific papers on these types of rock "Pillow Talk" and "Between the Sheets"! The scientific journal forced him to change the second title, which they considered too racy. In any case, beneath the sheeted dikes lie relics of the magma reservoir, forming a version of basalt called gabbro, which cooled and crystallized more slowly. The lower extent of the gabbro marks the base of the oceanic crust, making the crust typically seven to eight kilometers thick. Underneath this is the part of the ophiolite that gives it its name. This consists of remnants of Earth's mantle, forming the rock peridotite—named after the green gemstone peridot—which is composed mainly of the greenish mineral olivine and the dark mineral pyroxene. These minerals commonly undergo chemical alteration to form the "snake-like" scaly or platy mineral serpentine (*ophis* is Ancient Greek for snake, as is *serpens* in Latin). This happens when they are exposed to water under specific temperatures and pressures at depth, such as in subduction zones. This process transforms peridotite into serpentinite.

A FINAL RECKONING

Let us compare our newly acquired picture of the rocks beneath the ocean floor with those ancient volcanic rocks in the Barberton Greenstone Belt, which make up much of the Onverwacht Group. There is certainly no shortage of chert in the Onverwacht Group. But this is chert with a difference: it is not the fossilized remains of microscopic organisms. Instead, it is derived from a variety of other rocks that have been turned into chert, including volcanic ash, fragments of older chert, and various igneous rocks, including komatiites. For the moment, let us assume that these are the ancient equivalent of the top layer of the oceanic crust—later in this chapter I will talk more about how they came to be this way. Below the cherts, we certainly have lavas with the characteristic pillow shapes found on the ocean floor. Moving quickly to the third part of an ophiolite, there is abundant serpentinite in the greenstone belt.

It seems that we have all the ingredients of an ophiolite, and hence the remains of the ancient oceanic crust. In fact, Harald Furnes at the University of Bergen worked out the depth of this ocean. He made use of the fact that the pillow lavas in the Onverwacht Group sometimes contain the remains of gas bubbles, called vesicles. These formed as the lava erupted and the volatiles in the magma—water vapor and other gases—bubbled out in much the same way that gas bubbles form in boiling water. But because molten lava is much stickier than water, many of the gas bubbles remained trapped in the magma, leaving holes in the solidified rock. The size of these bubbles depends on the pressure of the ocean water into which lava is erupting. And the higher the pressure, the smaller the total volume of bubbles. The water pressure is determined by the depth of the ocean. For a significant proportion of the Onverwacht volcanic pile, the vesicles in basaltic pillow lavas point to ocean depths between 2 and 4 kilometers—typical depths for modern oceans, but certainly hundreds of meters.

It seems all wrapped up—or, as the English say, "done and dusted." But wait! There is a problem here. Remember those sheets? Imagine checking into a hotel and finding the bed unmade, with no sheets. The same applies to the Onverwacht Group. There are no sheeted dikes. This may seem a minor detail, but it certainly is not, as sheeted dikes are widely considered the dominant signature of seafloor spreading in the oceanic crust. When tectonic plates move apart, they create a gap that must be filled. At the spreading center, this gap is filled by vertical dikes of basalt. Each dike is no more than a meter or so wide, so at typical rates of plate motion—a few tens of millimeters per year—this would fill several decades' worth of gap. Repeated episodes of this process create ever-expanding walls of adjoining dikes, like continually inserting extra cards into a deck laid on its side—these are the sheeted dikes.

THROWING DOWN THE GAUNTLET

Maarten de Wit was very familiar with ophiolites—much of his early geological career had been devoted to studying the ophiolites in Newfoundland and Patagonia. And he had by now spent several years mapping the rocks in the Barberton Greenstone Belt—particularly the volcanic rocks in the Onverwacht Group—thinking long and hard about the apparent absence of sheeted dikes. He came up with a solution that was stunning in its simplicity. Here, I must reveal something that I have so far left unsaid. Geologists are so used to seeing rocks where the strata have been tilted by geological forces that they tend to automatically adjust for this by tilting the layers back to the horizontal plane in their mind's eye—from then on, all further discussion assumes this "correction." In the case of the volcanic rocks in the Onverwacht Group, the layers today are almost *vertical*—they have been tipped up on end, or sometimes even slightly beyond the vertical. This means that the once-horizontal flows of massive komatiitic lava now stand as vertical walls of rock—a bit like the missing sheeted dikes.

Could these so-called lava flows be the dikes? Maarten's idea went even further. He thought of the Onverwacht Group as the consequence of a giant hot intrusion of mantle, which had risen toward the seafloor at a seafloor spreading center. As it rose, it began to melt, and the magma pooled at depth in large magma chambers, while also making its way to the surface along vertical channels, pouring out onto the seabed as eruptions of pillow lavas. The sediment on the seafloor became caught up in all this, twisted around and sandwiched between flows, dikes and deeper intrusions. Maarten called it the Jamestown Ophiolite, with the use of "Jamestown" marking a deliberate break with prevailing thought. He wanted to return to an older subdivision of the "Swaziland System," defined by Arthur Hall on his 1918 geological map and named after the township where the first gold was found in the area. Maarten had, as it were, thrown down the geological gauntlet when he published this new vision in 1987. It caused an uproar among geologists worldwide, many of whom thought they knew all there was to know about komatiites—both in this region and elsewhere in the world. Given the significance komatiites were assuming in our understanding of early Earth, this was far more than a storm in a tea cup.

Maarten was the most scrupulous scientist I have ever met. He based his ideas on detailed observations of rocks in the field, and he had an uncanny habit of coming up with startling new discoveries. But he was also very conscious that he could be wrong. I remember him telling me on more than one occasion that the mark of a true scientist is a willingness to change long-held ideas if shown a new observation, even if your scientific career has been built upon those very ideas. This is a tall order for any scientist, and the history of science is strewn with great scientists who have refused to abandon wrong ideas. I think Maarten felt that science is about confronting and making sense of the natural world, unbiased by what you or other scientists used to think. In any case, having thrown down the gauntlet, there was bound to be a challenge.

The challenge came from one of Maarten's own students, who had been given the task of sorting out dikes from lava flows. Jesse Dann crawled over the river pavements along the Komati River and its tributaries, closely scrutinizing every feature of the famous komatiites. He made exquisitely detailed drawings and maps of the rocks, picking out contacts between adjacent dikes or flows, vestiges of sediment, and smaller features such as vesicles and individual crystals, as well as their relationship to the pillow lavas. He compared all this to modern lava flows. His verdict was unequivocal. The komatiites were lava flows, not dikes. The evidence was overwhelming. I remember Maarten telling me this in 2005, when I stayed with him in Cape Town.

LOOKING IN THE WRONG PLACE

Maarten lived up to his personal philosophy. He accepted Jesse's conclusions. But this is not the end of the story—far from it. First, there is the matter of scale. The komatiites form part of the bedrock of the Makhonjwa Mountains in a region that is at most 50 kilometers wide. Because the lavas are turned up on end, we do not know their original horizontal extent in the other direction, but not more than about 10 kilometers is preserved. A region this size would be a mere dot in comparison to the Pacific Ocean. And it is smaller than a single shield volcano today, such as Mount Etna, or the island of Hawaii. There is no guarantee that what we have preserved is typical of the rest of the ocean floor at the time when komatiite lavas erupted.

Indeed, as the crust beneath the present-day ocean floor is explored in greater detail, it is becoming clear that it does vary in nature. In the Atlantic Ocean near Portugal, there is no oceanic crust at all, and the mantle rocks come all the way up to the surface. The presence of sheeted dikes—rather than just widely spaced dikes—is not guaranteed either, as this is now known to depend on the interplay between volcanic activity and the movement of the tectonic plates. Their absence in the Onverwacht Group does not mean that there was no seafloor spreading. Also, it is possible that there are sheeted dikes that are simply not exposed at the surface today.

Finally, it may be that the oceanic crust that ended up on land, preserved as ophiolite, was not the typical oceanic crust found on early Earth.

Maarten had shown with his careful field observations that even if the komatiites are lava flows, there are other ultramafic rocks that are not. The lavas are engulfed in a sea of intrusive ultramafic and mafic rocks, including plenty of dikes—although not tightly stacked together to form sheeted dikes—and these rocks rise from the underlying mantle, much as in Maarten's original vision of the Jamestown Ophiolite. All this would have added to the crust, allowing it to grow and spread. It is likely, however, that the larger amounts of melting inside early Earth, as indicated by the eruption of komatiites and komatiitic basalts, would have built up a thicker oceanic crust. The crust beneath the oceans today is 7 to 8 kilometers thick. It has been calculated that early Earth's oceanic crust was at least twice as thick as this. As a final coda to all this, Maarten, working with his close colleague Harald Furnes, did find ancient sheeted dikes—not in the Barberton Greenstone Belt, but in the even older volcanic rocks at Isua in Greenland. So, it seems clear that seafloor spreading was taking place at this time, most likely as part of plate tectonics.

HORSES FOR COURSES

Today, the deep workings of Earth involve both plate tectonics and mantle plumes. It is the interaction of these two phenomena that create much of our planet's surface. Where plumes rise beneath the tectonic plates, hotspots of volcanic activity occur. As the plates move over these hotspots, trails of volcanoes form, creating long volcanic chains. These chains are clearly visible in all the world's oceans, although most of the volcanoes are underwater.

Sometimes, an unusually large plume develops—a superplume—and the amount of volcanic rock is so great that it builds up large portions of the seafloor into what are called oceanic plateaus. The largest oceanic plateau known on Earth came into existence during a brief period of intense volcanism about 120 million years ago. Subsequently, it was broken up into fragments by the motion of the tectonic plates, scattered to all points of the compass—south to New Zealand, and north and east of New

Guinea, forming the Ontong-Java, Manihiki and Hikurangi submarine plateau remnants. If the fragments are put back together, like a gigantic jigsaw puzzle, then they make up an oceanic plateau which extended over a thousand kilometers in all directions, underlain by oceanic crust up to five times its usual thickness.

Superplumes may be the closest modern analogy to plumes in the hotter mantle of early Earth, and geologists often invoke them as an explanation for the eruption of komatiites and komatiitic basalts at that time. In this interpretation, the creation of the thick pile of volcanic lavas in the Onverwacht Group—and of greenstone belts in general—would be analogous to the development of oceanic plateaus. Some of the lavas may still have erupted at the bottom of a deep ocean, but this removes any need for sheeted dikes.

I would argue that not all komatiites share the same origin. I use the term komatiite more loosely here, and in what follows, to refer to both true komatiites and komatiitic basalts. I should emphasize too that many of the volcanic rocks in the Onverwacht Group are *not* komatiites—there are plenty of typical basalts too, as well as volcanic and intrusive rocks that are richer in silica. It also turns out that these volcanic eruptions span a much

Sea Floor Spreading, Subduction and Plumes

ONVERWACHT times (3.5 to 3.3 Ga)

Crust of komatiites, komatiitic basalts & basalts

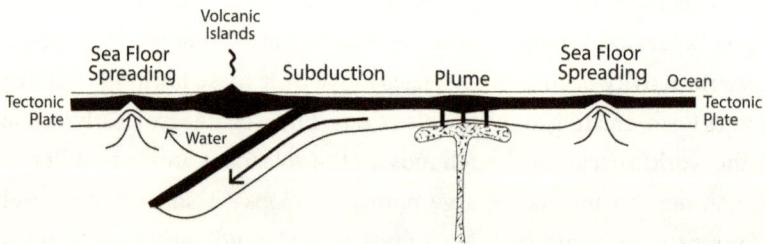

FIGURE 3.7 The volcanic rocks preserved in the Onverwacht Group are part of the crust beneath an ancient ocean. This hypothetical schematic diagram shows how they could result from volcanic activity driven by sea floor spreading and subduction, as part of plate tectonics, and also the rise of underlying mantle plumes. (Illustration credit: Simon Lamb)

longer period than was once thought. Using those zircon time capsules—introduced in chapter 2 and found in volcanic rocks or even in some of the cherts—it has been possible to pin down their ages more precisely. This work shows that the komatiites preserved in the Barberton Greenstone Belt represent about 250 million years of geological time.

The oldest well-preserved and least-altered examples of komatiites and basalts are those originally discovered by the Viljoen twins, making up what are called the Komati and Hooggenoeg Formations. These date to between approximately 3.48 and 3.45 billion years ago, a span of about 30 million years—this would be when they erupted on the seafloor. At some point in the following 120 million years, there was the eruption of a relatively thin sequence of undated komatiitic and basaltic volcanic rocks, known as the Kromberg Formation. Finally, in the subsequent 30 million years, a younger sequence of komatiite and basalt lavas accumulated (referred to as the Mendon Formation) until roughly 3.3 billion years ago. Given that individual eruptions of lava must have occurred during brief periods of volcanic activity, these dates indicate long periods of geological time that are unaccounted for in the rock record here. To put all this in perspective, consider that the great ranges of the Andes, the Himalayas, and the Tibetan Plateau, have only risen in the last 50 million years; the surface of the planet can change profoundly in a few tens of millions of years, and certainly in a hundred million years.

I think these timings are strong evidence for plate tectonics at this time, as both the duration of the volcanic episodes and the missing geological record are similar to the typical life of an ocean, as suggested by the age of today's seafloor. The long gaps do not mean that nothing was happening, just that the region had become part of the interior of an oceanic plate, whereas the geological action is mainly at the edges. But from time to time, when a plate moved over an underlying mantle plume, there would be a renewed flare-up of volcanic activity with prolific eruptions of both basalt and komatiite. However, given the ongoing controversy over the origin of the komatiites in the Barberton Greenstone Belt, there is still no consensus about which parts of the volcanic sequence are the product of mantle plumes and which are due to plate tectonics, such as where seafloor spreading occurs above a subduction zone. Maarten de Wit and Harald Furnes lean toward the latter explanation for most of the komatiites, based on their study of the rocks' nature and chemical composition, as I have already described. In the remainder of this book, I will argue that plate tectonics was alive and well at this time.

What almost all komatiites have in common is that they erupted in the first half of Earth's history when the mantle was, on average, hotter than today, and possibly had more water locked up inside, too. Mantle plumes and plate tectonics have acted together to drive the subsequent cooling, in effect stirring up Earth's interior by bringing hotter rocks to the surface and subducting colder ones, eventually ending the main age of komatiite eruptions. I will return to these ideas in chapter 7.

- § -

DEEP DIVING

What would it be like to explore the deep seafloor of early Earth? An important clue lies in a journey to the bottom of one of today's oceans. However, getting there in person is a daunting undertaking, requiring a dive in a tiny, ultra-strong submarine made of titanium alloy, designed to withstand pressures hundreds of times greater than at the surface. At these depths, there is no natural light; the ocean floor becomes visible only in the powerful beams of the submarine's searchlights. These dives have become increasingly rare because of their high costs and danger, largely replaced by remotely operated vehicles controlled from a mother ship stationed nearby. Either way, what would you see?

I suspect that most people's vision of the deep seafloor is that shown in the movie *Titanic*, based on the exploration of the wreck by Robert Ballard. The ship sits eerily on a flat and featureless landscape, with the occasional fish swimming in and out of portholes, and strange stalactite-like "rusticles" hanging from railings or the iron hull. The ocean where the *Titanic* sank is nearly 4,000 meters deep, and it landed on what oceanographers call the continental rise. Here, the deep seafloor begins to shallow, nearer the edge of the ocean and far from any submarine volcanic activity along the mid-ocean ridge. The accumulation of sediment has had time to smooth out the seabed. As you move east you reach the deeper abyssal plains, which are also smooth but with a thinner blanket of sediment. But if you go about 1,700 kilometers east of the *Titanic* wreck, right in the middle of the Atlantic, you will encounter a very different seafloor topography. Here, you are diving at the crest of the mid-ocean ridge in water that is about two to three kilometers deep. Now the seafloor has deep chasms, narrow ridges and towering volcanic mountains. Bare rock has the characteristic twisted and bulbous shapes of underwater volcanic eruptions. Manned submarines have to be piloted very carefully through this maze of craggy hills, rather like a slow-motion equivalent of those exhilarating low passes of small spacecraft in the Star Wars movies, as they skim through the canyons on the surface of some strange rocky planet.

It is clear that there is a wide range of scenery to choose from when exploring the seabed. The oceans of early Earth would have been no different. I imagine the seafloor during the early eruptions of the volcanic rocks in the Onverwacht Group to be similar to the rugged terrain of a modern mid-ocean ridge, although the hot and runny ancient komatiite lavas were capable of flowing several kilometers, smoothing out the terrain they covered. There were still the same bulbous pillow shapes where a magma of komatiitic basalt had burst out through cracks in the chilled carapace of the flow. But there were no shells or tests of dead organisms, just fine rock particles or volcanic ash—including layers made up of minute balls of accretionary lapilli that were washed in from shallower water—filling up crevices in the tops of the lava field.

Another danger lurked here. Everywhere, jets of hot water gushed out of cracks, sometimes clear and sometimes forming black or white clouds. Strange chimneys of iron and chert, meters high, dotted the seafloor like termite mounds—their remains can be seen in the rocks of the Onverwacht Group, either protruding from the chert layers, or as patches of heavily mineralized rock. Some of these have a greenish color, and others are black. They grew where silica and metals (or their chemical compounds) such as iron, copper, chrome, nickel, and gold, precipitated on the seafloor from the hot water. Maarten and his colleague Harald Furnes have even worked out the temperature of the water in the vents—it was superheated to about 200°C.

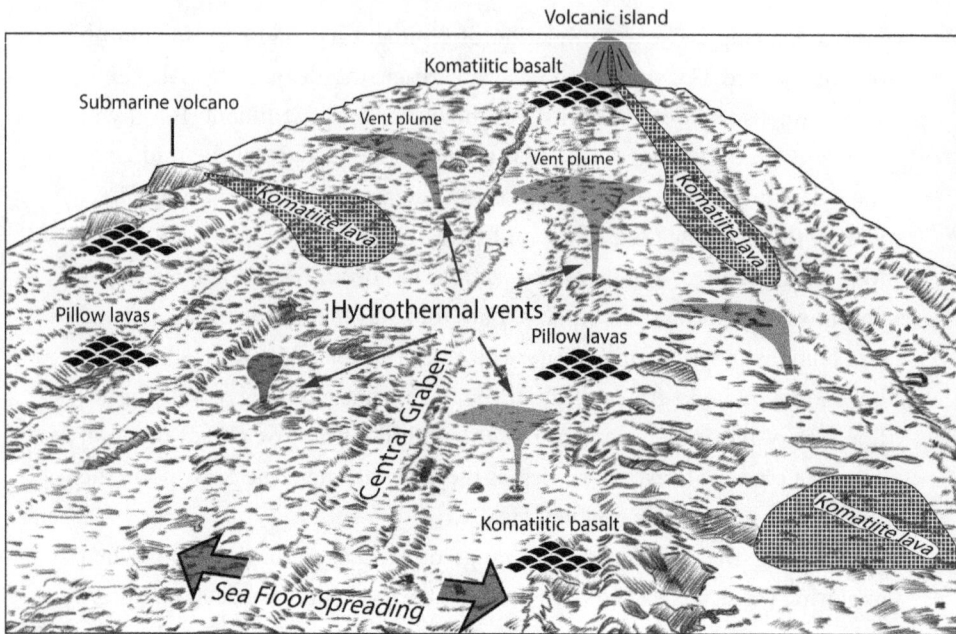

FIGURE 3.8 The seafloor near a modern seafloor spreading center may be our best guide to what the ocean floor looked like when the Onverwacht Group volcanic rocks were laid down, forming the basis of this reconstruction. A major difference, though, would be the extensive, runny flows of very hot komatiite lava, and thick piles of komatiitic basalt pillow lavas. The seafloor was likely dotted with many more hydrothermal vents, where superheated water gushed out, bringing up metals scavenged from deeper volcanic rocks. Like today, these vents supported life—although on early Earth, the organisms were single-celled microbes. (Drawing by Clara Maxwell Lamb)

The volcanic and sedimentary bedrock was reacting with the silica coming up in the vents, soaking it up like a sponge and slowly turning into chert—just as Maarten had shown me soon after I first met him. This gives the chert layers added significance, as their formation takes time. In other words, they mark long pauses between volcanic eruptions. This means that where more chert layers appear higher up in the volcanic pile of the Onverwacht Group, the frequency of eruptions was less. Taking the modern oceans as a guide, one could also be looking at pieces of seafloor that formed progressively farther from the main center of volcanic activity along the mid-ocean ridge. The seafloor was becoming smoother and more like the abyssal plains or continental rise. Interestingly, the reduction in the total volume of vesicles in the lavas suggests that the oceans were becoming deeper at this time as well—presumably because the seafloor had more time to cool and subside, much as in modern oceans.

LIFE GETS GOING

Hot vents are found today at the mid-ocean ridge or on the flanks of active volcanoes. They are home to thriving communities of worms, shrimp, and bacteria—for example, "Shrimp City" on the flanks of West Mata volcano. Was there any life around hot vents on early Earth? If so, this would be a likely place where life started in the first place, in an environment rich in minerals and energy. The answer, I think, is yes—though it consisted only of microscopic, bacteria-like organisms. When they died, their single cells were converted into rock by the aggressive chemical activity in the hot water. We know this because their remains have been found in the ancient cherts of the Barberton Greenstone Belt. These are some of the earliest known life forms, living 3.2 to 3.5 billion years ago.

The first discoveries were announced in the late 1960s in a slew of sensational papers by Elso Barghoorn, William Schopf, Albert Engel, and their collaborators, published in the prestigious American journal *Science*. These studies described what appeared to be fossilized microorganisms in the cherts of the Onverwacht Group, and the overlying Fig Tree Group. These authors scrutinized the cherts with high-powered optical and electron microscopes, identifying objects that had the typical sizes and shapes of modern bacteria, and sometimes contained carbon. A decade on, Andrew Knoll published a paper with Elso Barghoorn describing bacteria-like objects in these rocks that appeared to be in the process of dividing, in just the way modern bacteria reproduce, before becoming entombed. Since then, many more discoveries have revealed the wide variety of habitats that supported early life, both on the deep ocean floor and in shallow waters. I will talk more about this work in chapter 4.

I once collected a sample of chert from the Barberton Greenstone Belt in which fossilized remains of microorganisms were discovered. This took place on another occasion when I was with Maarten, in 1996, during the making of the BBC *Earth Story* series. We were standing near a low ridge of greenish-colored chert, part of the Onverwacht Group, and Maarten was telling me that this would be a likely place to find early signs of life. At that time, a colleague of Maarten, called Frances Westall, was developing techniques to find fossilized bacteria and other microorganisms by etching the rock with a strong acid to reveal the organic relics, then imaging the three-dimensional shapes with a high-powered scanning electron microscope. Maarten and I agreed that we should collect some chert for Frances. I asked Maarten what particular spot I should sample. 'Oh, anywhere," he casually remarked. I broke off the nearest bit of chert with my hammer and gave it to him. Some months later, Maarten reminded me about this sample: "You know, we found fossilised bacteria in it." It seemed much too easy. To me it suggests that even at this early stage, microorganisms must have existed in very large numbers. In fact, signs of early life have now been found not just in the cherts, but in amongst the volcanic rocks of the Onverwacht Group, clinging to the edges of individual pillow lavas and around the ubiquitous hydrothermal vents.

We have probably come as far as we can—given our present knowledge—in exploring the early oceans. The picture will certainly come into sharper focus as new clues are uncovered, both from the Barberton Greenstone Belt and from other greenstone belts around the world. Still, I can't help but be astonished by the detailed images we already have of the seafloor—and of the crust and mantle deep below it—during the early years of our planet. In the next chapter I will explore what happened at the edges of these oceans, where the land first met the sea.

- § -

BORDERS, LEOPARDS, HIPPOS, AND OTHER THINGS

The finale of my first visit to southern Africa in 1980 was a geological excursion to another greenstone belt, this time in Zimbabwe, about 600 kilometers north of my study area. The independence of Zimbabwe had only been declared in April, and so my visit was less than six months into the history of this new country. It had just emerged from a violent civil war that had pitted white settlers against the indigenous population, and the wounds of this struggle were still very raw, as I was to discover.

I traveled by car with my contact from the Swaziland Geological Survey. We entered South Africa at the northern border crossing, then headed up to the Zimbabwe border at Beitbridge. The border here follows the Limpopo River, and Beitbridge is quite literally the bridge that crosses it—a huge box-girder structure that, at the time, carried both the road and the railway line. It is named after Alfred Beit, one of the founders of the De Beers mining company, along with Cecil Rhodes. The border was still managed by the former white Rhodesian police, and they were very, very angry about their perceived treatment by the British government.

You may recall that Rhodesia was a British colony that declared independence—through a Unilateral Declaration of Independence (UDI)—as a white-run state in 1965. This action led to fifteen years of guerrilla war between the indigenous population and the white Rhodesian regime. UDI was not recognized by the international community, so under international law, Rhodesia remained a British colony—and, as far as the British government was concerned, an embarrassment. In the late 1970s, when it became clear that the situation in Rhodesia was rapidly becoming untenable, the British government negotiated a settlement. This involved the establishment of a Black majority government, eventually led by Robert Mugabe. Some Rhodesians felt ignored in the settlement, after fighting so hard to maintain their state, and they had become extremely bitter about what they saw as a betrayal. Others embraced the new country wholeheartedly, willingly giving up their right to British citizenship and committing themselves to building a better place for all.

I was made aware of some of this at the Beitbridge border post by a white border official as soon as he saw my British passport. As a preamble, you should know that at this time, border crossings anywhere in southern Africa could be stressful. A South African entry stamp in your passport could render your passport useless for travel in other parts of the world, as part of global protest against the ghastly Apartheid regime. But South Africa was desperate for foreign tourists, so it was generally accepted that the entry and exit stamps could be put on a loose piece of paper slipped into your passport, at the discretion of the border police. On a number of occasions, I found Afrikaner policemen very reluctant to do this; they saw it as an insult to their country. It also created the nightmare scenario of misplacing the piece of paper, at which point you immediately became an illegal visitor. Given all this, I thought I was prepared for the Zimbabwean border.

The white border official looked at me, and I could see from the expression on his face that he was not happy. He started off by asking me why he should let me into Zimbabwe. I explained the geological excursion to Belingwe. He replied by saying that "you British have betrayed our country." Then he told me that many of his friends had been killed in the war. Finally, he dismissed me, refusing to deal with me any further. I was stunned and had no idea what to do. Although I was fairly sure he didn't have any legal grounds for preventing me from entering the country, he was certainly not going to make it easy. He made me wait for about an hour as a stream of local people moved through the border post. When he felt that he had pushed me as far as he dared, he waved me to come forward, grabbed my passport and stamped it, giving the clear impression that I was not welcome. I stepped out of the border post dripping with sweat, and frightened.

Outside I was confronted by a strange sight. Parked along the road were an array of weird-looking vehicles with angular bodies mounted high off their chassis. Most of them were modified Land Rovers or trucks, and I learned later that each design had a particular name—leopard is the one that sticks in my mind—designed to deflect the blast of a road mine. As recently as a few months ago, it was still too dangerous to drive through much of what was southern Rhodesia on your own, and all traffic traveled in convoys. The strange vehicles at Beitbridge were the military vehicles that escorted the civilian convoys. I was told that the convoy would be assembled with a front and back escort vehicle, and would leave Beitbridge on the main road north, going as fast as possible without stopping until it reached the next fortified town. Any car that broke down would simply be abandoned, and its passengers quickly picked up by the rear escort vehicle.

There were no convoys now, but we were told that there was civil unrest in the region so we should still travel as fast as we could, not stopping under any circumstances, until we reached the next township. All this added a level of fear to the journey. When we reached the township, we found that it was surrounded by a fence made of huge coils of barbed wire, with a gate across the road. Inside, the main street was lined with a wall of sandbags, screening all doors and windows. I began to appreciate some of the realities of this very nasty guerrilla war. We finally arrived in the small town of Shabani (Zvishavane today), on the edge of the Belingwe Greenstone Belt, where we checked into the Nilton hotel—I could never decide whether this was a genuine local name or a playful allusion to the luxury Hilton hotel chain. Anyhow, it had its own unique charms.

Gathered at the hotel were the luminaries of the old Rhodesian geological community—scientists who had made their names studying the oldest rocks here. These are the names I remember: Jim Wilson, Tony Martin, John Orpen, and Euan Nisbet. It was an important excursion for them because it was the first time that they had been able to get back into the area after many years—it had been too dangerous during the civil war. And some of the big names in South African geology were here for this reason too: Carl Anhaeusser, Richard Viljoen, and Maarten de Wit. The Belingwe Greenstone Belt is about 700 million years younger than the Barberton Greenstone Belt, but many of the rocks are very similar. Life was well established on Earth by now, and it had left its mark in some beautiful mounds of limestone called stromatolites. You can see living stromatolites today along the coast of Western Australia, in Shark Bay. Here, in a subdued landscape of tidal flats—very similar to the environment in which the Belingwe stromatolites once thrived—time seems to have stood still.

My memories of the Belingwe excursion are mainly of intense arguments about the significance of virtually every outcrop of rock, in which nobody seemed to agree on anything—except the stromatolites, I think. But it was an energizing experience for me as a young geologist. We had some moments of alarm when it seemed that there might still be armed guerrillas in the region—our guides had rifles to protect us, but it proved unnecessary in the end. However, the excursion very nearly ended in disaster for me. One sunny and cool morning we visited some pristine examples of volcanic rock with the distinctive bulbous pillow shape, exposed near the banks of a river lined with thick reed beds. Rather negligently, I wandered toward the river and hit the first rock I came across with my hammer. A shard of rock flew straight up into my eye. The pain was excruciating, and I was half-blinded. A moment later, an enraged female hippopotamus, mouth open to reveal her tusklike teeth, charged out of the reed beds straight toward me. Everybody started shouting. I ran away, not really knowing where I was going. I was fortunate: the hippo had made her point and retreated back to her wallowing spot in the river. I was told later that she had her calf with her, and I had gotten too close.

By this time, I was in agony and my eye was bleeding. I was rushed to the nearest mine hospital where a doctor examined my eye using a sort of microscope. He said that the eyeball had been cut by the shard of rock, but he was pretty sure it hadn't pierced the orb. He gave me anesthetic eye drops to reduce the pain and advised me to see an eye specialist when I got back to England. I was very conscious of the risk of eye damage, because one of the other participants on the field excursion had been blinded in one eye in a similar accident. Fortunately, there was no permanent damage to my eye. But all this concern seemed minor compared to the excitement of the previous few months during my first visit to southern Africa.

4

On the Edge

Where the sea meets the land: the life and death of the first coastlines

So far in our journey on the surface of the young planet we have been exploring the bed of an ancient ocean. This, you will recall, is preserved in the Onverwacht Group of the Barberton Greenstone Belt and forms part of the bedrock of the Makhonjwa Mountains. Our perspective of Earth at this time is similar to that of a ship—or perhaps a submarine—far from land. The land is clearly somewhere, but we can't see it yet. Its existence is indicated by those layers of chert, some of which are the altered remains of silt and mud that were washed into the ocean by rivers, or blown by the wind from some dusty desert, and then deposited on the seafloor. Also, we can see volcanic ash in the chert, forming balls of accretionary lapilli which must have erupted from a volcano into the atmosphere, rising as a cloud before raining down again on land and sea. The ash presumably came from isolated volcanic islands or a great chain of volcanoes forming an island arc. There are clearly whole new worlds to be discovered at the edges of this ancient ocean. In this chapter,

I want to take you to these places on early Earth, where the sea meets the land. To set the scene, let us return once more to the ocean, but this time aboard the New Zealand research vessel *Tangaroa*, near the end of a cruise to the Southern Ocean, south of New Zealand.

OUT AT SEA

My abiding memory is an endless feeling of seasickness, caused by the constant motion as the prow of the ship plunged heavily into the swell, sending wave after wave of white water crashing over the hull. From the ship's bridge, it was the horizon that appeared to rise and fall alarmingly as the *Tangaroa* forged ahead. Violent gusts of wind, reaching over 100 kilometers per hour, would hit the ship sideways and force the ship to heel over. This sent water streaming across the deck toward the scuppers where it was blown into a white spume. Overhead, albatrosses seemed oblivious to all this turbulence, expertly managing the strong air currents by tipping their huge wings as they followed the ship through the storm.

It was March 2009, and the *Tangaroa* was laboriously returning to New Zealand after completing its research tasks. Instruments from deep ocean moorings had been retrieved as part of a long-term project to monitor changes in the Southern Ocean. As far as the scientists on board were concerned, I was here to learn about how this research was adding to our knowledge of the impact of global climate change on the oceans. But I had a secret motive of my own. I had been thinking for so many years about the evidence for long-vanished oceans in the geological past, I was beginning to feel I should see for myself what it is really like to be on the high seas. This opportunity seemed too good to miss. I soon had plenty of experience of the extreme conditions at the sea surface, tossed about in rough water and buffeted by the wind. But one activity during the voyage gave me another dimension to the ocean, really bringing home just how much water there was below me, and what lay at the bottom. This was the collection of sediment samples, which involved lowering a strange tripod contraption with a set of downward-pointing tubes into the water. It took over three hours to reach the bottom, as kilometers of cable unwound from the winch drum.

The weight of the tripod would then force the tubes into the muddy sea floor, filling them with slugs of sediment.

Once the tripod was back on board, I felt a sense of wonder as I peered at this ordinary-looking gray mud encased in battered plastic tubes, marveling that it had come from a depth of over four kilometers (2.5 miles). The mud must have slowly accumulated through the settling of dead marine organisms, along with detritus likely washed in or blown from onshore New Zealand, the nearest land. It struck me that I was looking at the modern equivalent of some of those layers of chert in the Onverwacht Group before they had been turned into rock. Yet the Onverwacht cherts had formed at the other end of our planet's history, nearly 3.5 billion years ago, and were now exposed in the Makhonjwa Mountains of southern Africa, visible to any geologist who made the effort to visit them.

My time at sea showed me much more than just what shipboard research was like. I also had the opportunity to view the exposed ocean-facing coastlines of eastern New Zealand from what was a new angle for me, looking landward from the sea. As we sailed closer to the east coast, I spent much of my time talking to the captain, watching him plot the ship's track on a map. Our course was almost identical to that taken by Captain Cook, nearly 250 years earlier, aboard HMS *Endeavour* on his first voyage of discovery to New Zealand. Cook's view of the world was that of a seaman, seeing land from the perspective of the ocean. His first sightings of land were of that dangerous 'edge' where ocean waves pounded the rocks, creating what seemed from out to sea to be a long white line of frothing surf, spray and breakers. Like Cook, the first sight I had of land was Young Nick's Head, named after a cabin boy on the *Endeavour*, and now forming a ghostly gray ridge on the horizon. And in front of it, extending as far as the eye could see, was the great line of white breakers. It was impossible to hear the sound of the surf in the screaming wind, but there are many descriptions of this ominous roar in the journals written by Cook and the scientists on board the *Endeavour*.

A few kilometers out from the turbulent reefs, the *Tangaroa* turned parallel to the shore and sailed northward. I spent hours observing through binoculars the high cliffs and coastal platforms along this coast, many of which are difficult to access from the land. There were few roads that penetrated the rugged hills behind, and only very occasionally could I see signs of settlements. I began to understand how Cook and his crew might have felt as they coasted these unknown shores, naming headlands based on features that were only visible from the sea. Eventually, we rounded both East Cape and the Coromandel Peninsula and passed through the heads of Waitematā Harbor into the relative safety of a large inlet, docking at the naval base. It was strange to be on land again, and I found myself instinctively bracing for the roll of the ship as I walked along the wharf and looked back at the vessel that had brought us safely through the dangers of the Southern Ocean.

MUD POOLS

With these graphic images of the watery world of the ocean and its edges in mind, let me return to those ancient rocks in the Barberton Greenstone Belt. If we imagine these rocks as a thick pile of volcanic and sedimentary layers, stacked on top of each other, then there are several places in this stack where we effectively make landfall, leaving the ocean and crossing over to dry land. I will just focus on the top part of the stack. Here, we have fast forwarded about 150 million years from the time of that early ocean floor, which I described in chapter 3, to roughly 3.3 billion years ago.

It was Maarten de Wit who found unusual evidence for what the edge of the ocean was like at this time. As he told it to me, it was while he was eating his sandwiches during a lunch break in the field that he noticed something odd about the rocky outcrop where he was sitting. It was one of those cherts, made up mainly of layers of quartz—each layer a few millimeters to centimeters thick—alternating in color between red, white, gray, and black. The red layers are called jasper, which is just quartz stained by a rust of iron oxide. The black, gray and white layers contain quartz crystals of different sizes or different amounts of impurities—the rock appears dark gray or black if the crystals are very small, and white if they are large. The black layers can also be pure iron oxide, in the form of magnetite or hematite. Typically, I would call this type of rock a banded ferruginous (that is, iron-rich) chert—if it is exceptionally rich in iron,

FIGURE 4.1 A typical banded ferruginous chert layer in the Fig Tree Group, made up of colorful layers of red iron-rich chert (jasper), occasional dark layers rich in magnetite and hematite, and intervening black and white iron-poor chert layers. Past earth movements have twisted the layers so they now are mostly vertical. (Photo credit: Simon Lamb)

it is usually called banded iron formation. There is plenty of it in the Makhonjwa Mountains.

You need to be able to visualize his lunch spot in among the outcrops of banded ferruginous cherts in order to understand Maarten's discovery. These cherts tend to form blocky outcrops, with many planar rock faces where the rock has split along individual layers. The rock faces are therefore the original tops or bottoms of these layers, but they now form vertical walls because the layers have been turned up on end as a consequence of the huge geological forces that squeezed the greenstone belt at a later stage of its history—I will talk more about this in the next chapter. From his seat, Maarten had a clear view of numerous rock faces and was struck by a strange, swirly pattern of lines etched into their planar surfaces by slight differences in the depth of weathering. It seemed as if he were looking at parts of a mural or some form of natural graffiti, repeated multiple times, as many of the surfaces displayed the same pattern. The swirls formed nests of concentric lines, abruptly truncated by neighboring nests. Imagine raking a smooth patch of sand in a series of curved sweeps that leave concentric parallel grooves in the sand. If you shifted your rake from time to time to reposition your sweeps, also raking the edges of earlier ones, you would have something similar to the swirly patterns Maarten had spotted.

Rather than just forget about what he had seen, Maarten puzzled over it, and it stayed lodged in his mind. This was lucky, because several months later he was invited to a conference in New Zealand, and included with the conference was a field excursion to visit the volcanically active region of North Island, near the town of Rotorua. An essential part of any trip like this is doing what most tourists who come to New Zealand also do—visiting the geysers and bubbling mud pools. Here, boiling ground water rises forcefully to the surface from a depth of several kilometers. Where this happens in the mud pools, the escaping gas forms bubbles that churn the surface of the mud, sending circular ripples outward as expanding concentric ridges of liquid mud. Eventually, the location of the bubble shifts, and older concentric rings are disrupted by newer ones, producing a distinctive, intersecting swirly pattern.

Maarten's visit to New Zealand was one of those rare moments of scientific revelation. He immediately realized that the patterns in the bubbling mud pools were identical to the ones he had seen in the layers

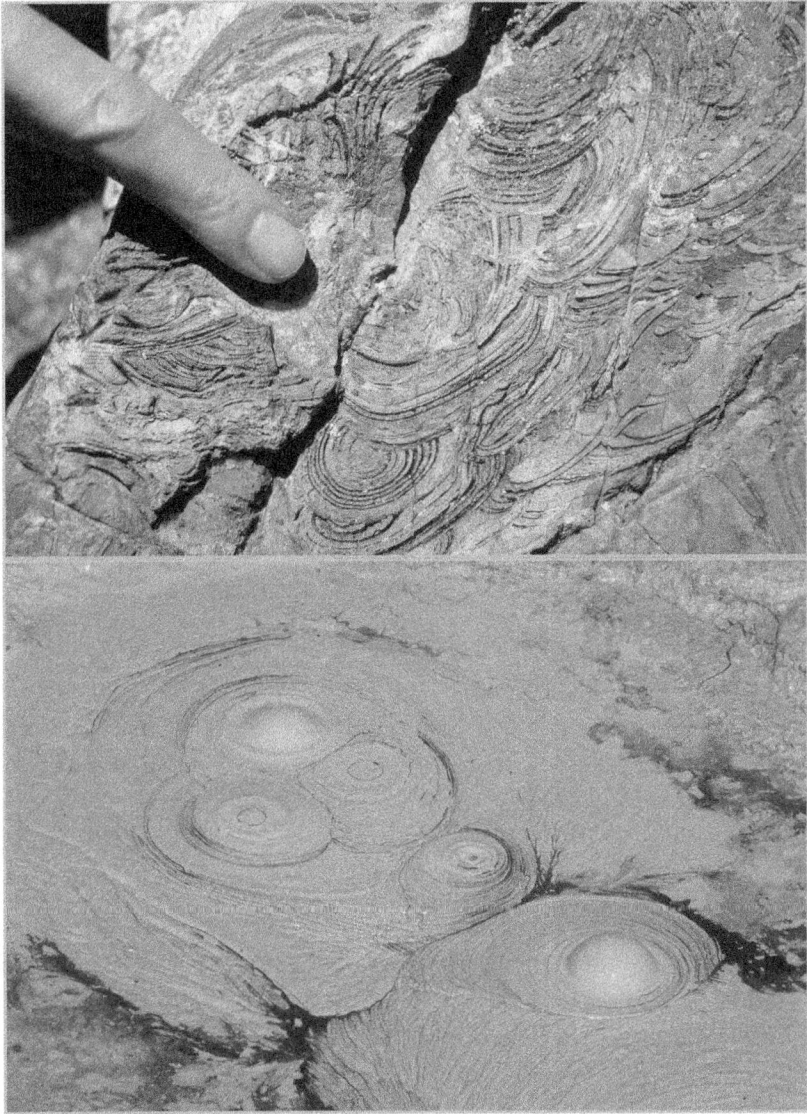

FIGURE 4.2 Above: Maarten de Wit's finger points out his discovery of possible fossilized mud pools, preserved as concentric swirly patterns in the layers of ferruginous (that is, iron-rich) chert found near the transition between the Onverwacht and Fig Tree Groups in the Barberton Greenstone Belt. Below: These features are remarkably similar in size and shape to the bubbling mud pools in modern geothermal regions, such as Rotorua in North Island New Zealand and, shown here, the volcanically active Andes of South America. (Photo credit: Simon Lamb)

of the banded ferruginous chert a few months ago in South Africa. Not only were the shapes of the swirls the same, but their sizes were too. There was no doubt in Maarten's mind that he had eaten his lunch during his African fieldwork on the fossilized remains of bubbling mud pools. But these mud-pools were well over three billions years old, compared to the ones that were gurgling before his very eyes in New Zealand!

Maarten, rather unsurprisingly, called the swirly features in the cherts "mud pool structures," and he soon found these at many other places in the greenstone belt. More fundamentally, their presence indicates an environment quite different from the deep ocean we explored when looking at the Onverwacht Group. Not all the chert layers in the Makhonjwa Mountains were laid down on the deep seafloor—as pointed out earlier on in this book, they are a wide variety of rocks in disguise. The presence of cherts with mud pool structures is a sure sign of dry land or very shallow water, where pools of bubbling mud could occasionally dry out and preserve traces of their activity. The mud pool structures exist in a particular part of the stack of rocks: round about the transition between underlying volcanic lava flows of the Onverwacht Group and an overlying pile of sedimentary rocks traditionally referred to as the Fig Tree Group. The Fig Tree Group is the focus of much of this chapter, but I have to say that when I first heard its name, I thought it was a strange one for a sequence of rocks—it conjures up in my mind an image that is more Mediterranean than African.

Not long after Maarten's discovery of the mud pool structures, I found them too, where I was working in the Malolotja Nature Reserve—and at the same level in the rock succession as that studied by Maarten in South Africa. In fact, my examples are just over the international border from where Maarten's student Isabelle Paris was looking at the rocks. She had found another extraordinary testament to the conditions around the mud pools. She had noticed that some layers of chert, when exposed so that one could see extensive layer surfaces, were full of pits. The pits ranged from a few millimeters to a centimeter across, and they looked just like the pits that hail can make in unconsolidated mud. I like to think these pits record a passing hailstorm, or perhaps they were cause by spatter from hot mud flung up by a particularly vigorous bubbling mud pool. Either way, they are an eloquent but very ephemeral relic of early Earth.

MOUNDS OF SLIME

Maarten's mud pool structures show that we have made that transition from ocean to land, but with no obvious sign of the intervening wave-pounded shoreline. I will return later to the question of where this turbulent shoreline might be, but in any case, I think we are still close to the ocean edge—perhaps in some protected lagoon or inlet. The mud pools, together with evidence for hydrothermal vents, show that there must have been magma at depth. And amongst the underlying rocks were old flows of lava, including those strange mantle-like volcanic rocks I described in the previous chapter. Sitting on one of these lava flows, Maarten found another curious rock. It formed chimney-like protuberances with fine internal layers, making the chimney look a bit like an upside-down stack of bowls. It was only when we were both on a field excursion in 1980—to look at the much younger (2.7 billion years old) Belingwe Greenstone Belt in Zimbabwe—that we realized what this rock might be. We were shown outcrops of limestone that had only recently been identified as the remains of microbial mounds known to geologists as "stromatolites" (from *stromae*, Ancient Greek for layers or laminations).

Stromatolites exist today, but they are restricted to a few unusual environments because they are very vulnerable to modern biological activity. The most famous place is Shark Bay in Western Australia, about seven hundred kilometers north of Perth. Here, you can walk around at low tide in water a meter deep or less, wandering through a strange uneven world of small mounds, like a field of molehills. Each mound occupies about the area of a cushion and is made of mud, silt, and a limestone crust, all covered in a green slime. It has been built up by successive drapes of microbes—technically cyanobacteria, but commonly called blue-green algae—that trap a sediment of mud and silt in their mat-like growth. The mound hardens because the microbes, as part of their metabolism (see below for further discussion), trigger the precipitation of calcium carbonate crystals (that is, limestone) from seawater, bonding the sediment particles together. The microbes become entombed in this cement, and a new mat of microbes grows on top. The whole process is repeated, and over time the mounds rise higher—sometimes reaching heights of a meter or so—with a characteristic

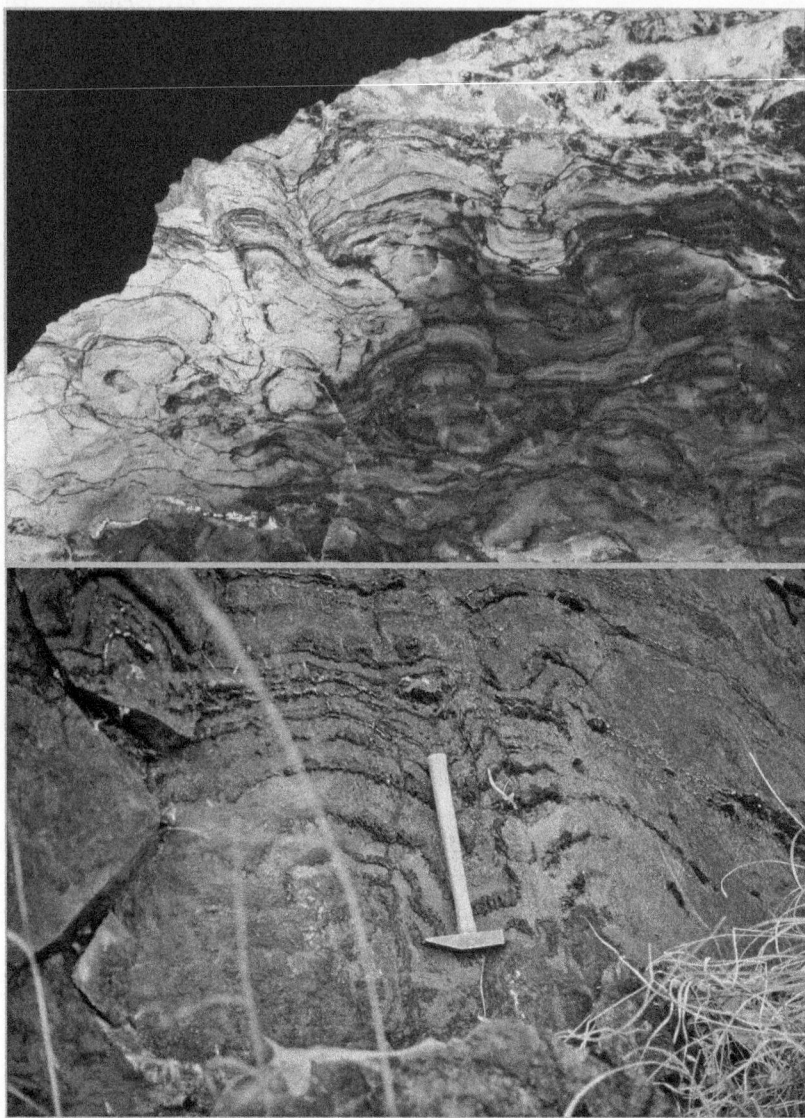

FIGURE 4.3 (*Top*) Possible remains of microbial slime that lived close to an ancient shoreline in shallow water, about 3.3 billion years old, preserved as dome-shaped mounds called stromatolites near the transition between the Fig Tree and Onverwacht Groups of the Barberton Greenstone Belt. The image shows a polished rock surface, now mainly chert and about 10 cm across. (Photograph courtesy of Cornel de Ronde); (*Bottom*) A clue to their origin comes from the shapes of typical stromatolites, such as those now preserved as limestone in the younger Belingwe Greenstone Belt in Zimbabwe, which thrived about 2.7 billion years ago. (Photo credit: Simon Lamb)

fine, dome-like internal lamination. It was this dome-like lamination that Maarten had spotted in the Barberton Greenstone Belt. However, the rock he was looking at had been altered by the corrosive effect of silica, so that it was now partly chert and not the original limestone.

Something similar to stromatolites can be found in rivers that flow through chalk terrain. These are not anchored to the riverbed but instead are loose, pebble-like concretions, rarely visible unless the sediment is disturbed. I remember as a child finding piles of them among the spoil heaps of a dredging operation to deepen the River Lea near my childhood home in England. They look like small meringues, ranging between the size of a marble and tennis ball, and have a fragile, crusty feel. But if you break them open, you will see numerous concentric layers a bit like those of an onion. I was fascinated by them, and I discovered that they are called oncoids. They form in the same way as stromatolites, but in this case the microbial slime grows on a loose pebble which is occasionally turned over in the river current. The growth of the microbes triggers precipitation of a limestone crust from the river water, adding a layer to the pebble. More microbes form on top of this, and so new layers are added, creating successive onion-like rings. Smaller versions of these crusts are called pisoids, or, if less than two millimeters across, ooids, although there is a debate about the role of microbes in the growth of the smaller balls, and they could also just be the result of direct precipitation from water. To my mind, if microbes were present on early Earth to build the stromatolite mounds, then it seems at least possible that loose balls like this could have formed as well. This might therefore be an explanation for some of the ubiquitous accretionary balls in the cherts of the Barberton Greenstone Belt, in addition to a volcanic origin (see more about these balls further on).

In 1986, a few years after Maarten's discovery, a group of American geologists headed by Gary Byerly from Louisiana State University confirmed his identification of stromatolites, finding them in other places in the same part of the stratigraphic pile. In the late 1990s, Maarten joined forces with Francis Westall to investigate more closely the organisms that created these features. I described in the previous chapter how I had broken off a sample of green chert for Maarten, back in 1996, in the hopes that it might contain ancient remains of life. Frances Westall made a detailed study of these and other cherts, collecting more of her own and examining them under the scanning electron microscope. She found that some of the cherts, forming layers round about the transition between the Onverwacht and Fig Tree Groups, had the telltale signs of organic molecules, called kerogens. The kerogens occurred as thin seams which she thought were the remains of microbial mats. Frances deduced from the detailed arrangement of these mats, combined with the relics of a particular crystal form of calcium carbonate called aragonite, that the microbes were capable of using solar energy to power what is today one of the most important chemical reactions of life: photosynthesis. But this is not the photosynthesis typical of modern microbes, such as cyanobacteria, algae, or the plant world.

I should now introduce the two versions of photosynthesis found in nature. Both depend on sunlight to drive their biochemistry. Microbes, such as cyanobacteria, use light energy to split carbon dioxide and water, producing oxygen and carbon compounds. This process, known as oxygenic photosynthesis, is carried out by organisms that can tolerate oxygen, including almost all plants. Almost all animals depend on the oxygen produced by oxygenic photosynthesis to breathe. A byproduct of this photosynthesis, when it is carried out by organisms that live in water, can be the local build-up of lime scale or "calcification" of the organism due to a precipitation of aragonite —hence the significance of the aragonite found by Frances in the cherts. In environments where there is no oxygen, there are organisms that can use light energy to split molecules such as hydrogen sulfide, releasing sulfur. This process, known as anoxygenic photosynthesis, can also drive calcification by aragonite. Thus, it is photosynthesis—either oxygenic or anoxygenic—that drives the buildup of limestone in the stromatolites' mounds. It used to be thought that the earliest microbes were capable of oxygenic photosynthesis, but it is now generally accepted that the anoxygenic version came first. Frances thought the photosynthesis indicated by her chert samples was anoxygenic. In addition to crystals of aragonite, she also found pyrite in amongst the kerogens, and pyrite is an iron sulfide that only forms in the absence of oxygen.

Frances concluded that the mats had been growing in clear and shallow water, close to land, so that sunlight could reach them; and that there was virtually no oxygen in the atmosphere and water, allowing the anoxygenic

FIGURE 4.4 Beautiful bladed crystals of barite (barium sulfate) in the Fig Tree Group, exhibiting what is sometimes called a cauliflower form. In modern springs, barite precipitation can be triggered by bacterial metabolic activity. The presence of this barite in the Barberton Greenstone Belt suggests thermal springs on land, most likely along the shores of active volcanoes. (Photograph courtesy of Cornel de Ronde)

bugs to thrive. Mounds of stromatolites built up locally where microbes, through photosynthesis, triggered large-scale precipitation of calcium carbonate, mainly as aragonite. Another consequence of this lack of oxygen was that iron remained unoxidized, allowing it to dissolve in water and be carried by rivers and ocean currents far from its original source. In contrast, oxidized iron is largely insoluble. Some anoxygenic microbes, however, as part of their photosynthesis, can turn this iron into insoluble iron oxides such as magnetite and hematite. This is a process known as photoferrotrophy, and it may explain the accumulation of the iron-rich sediments in the Fig Tree Group. A cloud of rust would form in the water, slowly settling to the bottom as reddish-brown mud, and eventually becoming ferruginous cherts or banded iron formation.

Another distinctive mineral found in these rocks is called barite, which forms thin seams amongst the layers of sandstones, shales and volcanic ash near the bottom part of the Fig Tree Group. Barite contains the elements barium—sometimes used as an inert material in medical imaging of the gastrointestinal tract—as well as sulfur and oxygen in the form of sulfate (a sulfur oxide). One possibility is that the barium was carried up to the surface in springs, precipitating as barite from a concentrated cocktail of chemicals and then crystallizing as beautiful cauliflower shapes. The sulfate may have originally come from volcanic gases spewed into the atmosphere that were then dissolved in the rain water that supplied the springs. However, microbes may also have played a role, as observed in modern thermal springs, by driving the precipitation of barite through their life-sustaining chemical reactions. In any case, the presence of this barite indicates the existence of dry land with pools and geothermal activity.

So what landforms should we envision for these rocks? I imagine volcanic islands with broad, fringing zones of very shallow water—rather like some modern-day Pacific islands—colonized by microbes. The microbes must have been brightly colored—pink or purple have been suggested (it seems that green as a biological color appeared later in Earth's history)—to help soak up sunlight for photosynthesis. They would have created a hummocky terrain, just like that at Shark Bay in Western Australia, slowly building up local mounds by trapping sediment particles and precipitating calcium carbonate. The line of crashing waves and white foaming water would have been much farther offshore, at the edges of fringing platforms where the water deepened to oceanic depths. But from a geological point of view, the modern world of stromatolites in Shark Bay is fast asleep, whereas the world of the early stromatolites was alive with volcanic activity. We must imagine volcanoes rising above the water, periodically erupting super-hot komatiitic lavas that flowed over the stromatolites in places, vaporizing the

Explosive basalt, komatiitic basalt and komatiite volcanoes

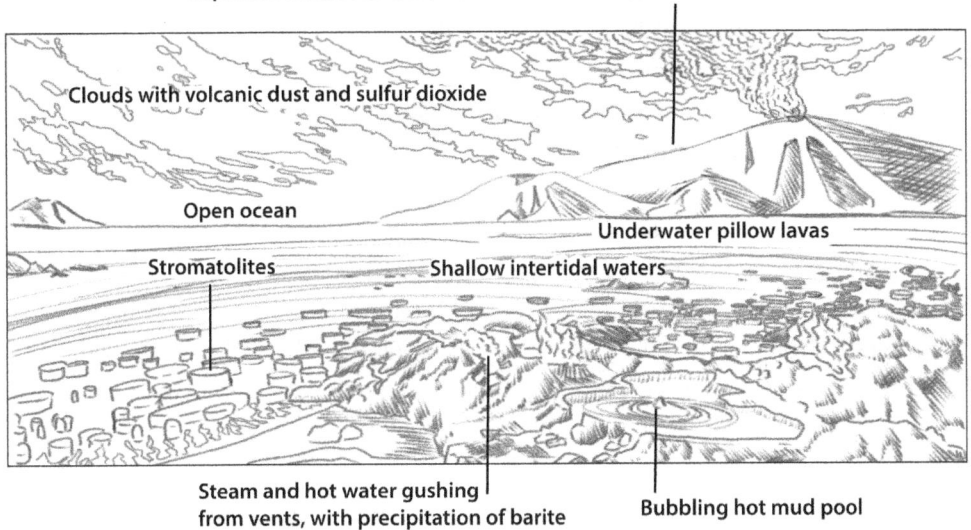

Clouds with volcanic dust and sulfur dioxide

Open ocean

Underwater pillow lavas

Stromatolites

Shallow intertidal waters

Steam and hot water gushing from vents, with precipitation of barite

Bubbling hot mud pool

FIGURE 4.5 A sketch reconstruction of volcanic islands during late Onverwacht and early Fig Tree times, roughly 3.3 billion years ago, based on features in the rock layers of the Barberton Greenstone Belt. The perspective is similar to that of the famous Smithsonian diorama of the Archean Eon by Peter Sawyer. This was a world dominated by volcanic activity, with bubbling mud pools and geysers, and volcanoes erupting superhot komatiites and basaltic komatiite lavas. Stromatolites and microbial mats thrived in shallow nearshore waters, surrounded by deep ocean. (Drawing by Clara Maxwell Lamb)

microbes. And, as Maarten had shown, there would be numerous pools of mud where volcanic gases and boiling water bubbled up, flinging out mud spatter and leaving behind the distinctive mud pool structures now preserved in the banded ferruginous cherts.

All this volcanic activity was not only creating the islands and surrounding seabed; it was also changing the very nature of the underlying rocks. I have already described how the cherts are really a wide range of rock types in disguise, and how the deep seafloor may be one of the factories for their production—the silica is ultimately derived from hot fluids expelled from magma at depth. Wherever these fluids come into contact with rocks—whether on the seafloor or on land—a chemical reaction is likely to take place that may eventually result in a chert. In a process that is still rather mysterious, the non-silica components of the original rock are exchanged for silica from the hot silica-rich water. This way, the rock can be converted to almost pure silica, with just traces of the other elements—in other words, a chert. But whatever is left behind causes a slight discoloration of the chert: the internal organization of the original rock—layers or crystals or grains—can still be made out from the traces of their original chemical components. The colorful cherts are a consequence of the particular elements left behind: iron makes it red, chrome makes it green. In the end, bubbling mud pools, stromatolites, and even komatiite lava flows all become different-colored cherts, with just enough clues left to work out what they originally were.

DON'T LOOK UP!

One particularly abundant type of chert from this time is full of small spheres, each several millimeters across. Some of these have all the features of accretionary lapilli, which form when volcanic ash clouds start to clump around raindrops, eventually raining down and blanketing the ground with small volcanic balls. Where I was working in Swaziland, cherts containing these balls were sometimes so weathered that they were crumbling into what looked like piles of olives or grapes, and I could pick out individual accretionary lapilli.

But some of the balls may have come about in a different way. I have already suggested that they may be a sort of "loose" stromatolite, created by microbial growth around a sand grain or small pebble. Don Lowe, at Stanford University, has proposed yet another origin: fallout from giant meteorite impacts. He based his conclusion on detailed observations of the balls—or spherules, as geologists tend to call them—by examining wafer-thin translucent slices of chert under the microscope. This way, it is possible to see the individual crystals and their shapes. Lowe identified particular balls that looked quite different from typical accretionary lapilli. Instead, they had the distinctive features of fragments of rock that have undergone shock melting due to extreme temperatures and pressures, forming glassy droplets. These are known as tektites from the Ancient Greek *tektos*, meaning molten. Glassy droplets can form in volcanic eruptions—perhaps melted by lightning strikes in the ash cloud—but they are more typically the product of the vast amount of thermal energy released by a large meteorite impact, which at the point of impact is enough to rapidly melt the surface of the Earth, flinging droplets of magma skywards. The droplets would solidify as tektites, falling out over a wide region.

Lowe found the spherules not only within distinct layers but also wedged into cracks within the underlying bedrock to the Fig Tree sediments. Large amounts of silica-rich fluid must have passed through these cracks, turning everything to chert so that the cracks appear as veins of chert cross-cutting the bedrock. Lowe suggested that the cracking was caused by shock waves radiating outward from ground zero of the meteorite impact site, literally shaking the bedrock apart. If this is correct, then they are direct evidence for the scale of the meteorite impact, although the actual site of the crater is unknown. Another explanation is that the cracks resulted from earthquakes, which shook the still-unconsolidated sediment, opening up cracks into which spherule beds subsequently fell. Or they might be a result of the shock waves from a violent volcanic explosion, or superheated water erupting in a bubbling geyser. In any case, they point to occasional episodes of extreme violence, disturbing the mud pools and microbial mats in the ancient shallow seas.

But something else was going on too. This is clear if one considers that much of the volcanic succession in the underlying Onverwacht Group may have formed on the deep ocean floor, at depths of two to four kilometers below sea level. Yet the mud pools and stromatolites were at or near sea level. And the region would soon return to deep water again—the Earth's surface seems to be acting like a yo-yo, bouncing up and down! Now is the time to embark on a new journey of discovery. Let me begin with something completely different: the sudden loss of telegraphic communication between Europe and North America in 1929. Bear with me and you will soon see why this is the key to understanding what the rocks in the Fig Tree Group are telling us about early Earth.

- § -

SHAKEN AND STIRRED

A magnitude 7.2 earthquake shook the Grand Banks region on the eastern seaboard of North America in 1929. It created a tsunami that killed twenty-eight people on the Burin Peninsula in Newfoundland and left a thousand or more homeless. Whole houses were washed out to sea; one of these was still floating when it was retrieved by local fishermen and towed back to shore. Twelve transatlantic telegraph cables, draped across the ocean bed and connecting North America with Europe, were cut. Surprisingly, not all of the cables were severed during the earthquake itself; instead, they broke over the following thirteen hours. The cause of this was a complete mystery.

It was not until the early 1950s that a convincing answer was found, when two American oceanographers—Bruce Heezen and Maurice Ewing—made detailed underwater surveys of the Grand Banks region in the vicinity of the cables that broke during the 1929 earthquake. To understand what they found, you need to know that the seafloor here forms a dramatic step at the edge of the continent. The shallow part, extending out to a depth of about 200 meters, is the continental shelf. At the shelf edge, it deepens dramatically down the continental slope to depths of about four kilometers. It then begins to flatten out again in the continental rise before reaching the abyssal plains at depths of five kilometers or so—the wreck of the *Titanic* lies on the continental rise off Newfoundland, about 500 kilometers away from the Grand Banks. What was surprising about the 1929 earthquake was that the telegraph cables had broken on *both* the steep continental slope and the deeper but flatter continental rise and abyssal plain, and it was the deeper ones that broke after the earthquake. Here Heezen and Ewing also found a large mound of sediment on the sea floor.

Heezen and Ewing realized that they were looking at the aftermath of a sudden and very powerful underwater avalanche of sediment, more technically called a submarine debris avalanche. It had flowed down the continental slope and over the rise, crossing the cables in the immediate aftermath of the 1929 earthquake. The cables lay in progressively deeper water, and so the debris avalanche had severed them one by one. By comparing the position of each break with the times when the cables had broken, Heezen and Ewing worked out the speed of this debris avalanche. In its early stages, it had travelled at nearly 100 km per hour, slowing down to about 20 km per hour after travelling over 700 km and reaching far into the ocean depths. This was clearly a major force of erosion, capable of shifting huge amounts of sediment over large distances in just a few hours. The most likely source of the debris avalanche was a landslide right on the edge of the continental shelf, triggered by the violent shaking of the 1929 earthquake.

Since Heezen and Ewing's original study, oceanographers have investigated submarine debris avalanches in great detail, both by studying the sea floor—not just on the Grand Bank, but also the steep edges of the continents elsewhere in the world's oceans—and by creating their own experimental examples in tanks of water. The debris avalanches are now recognized to also be a type of "turbidity current" in which a dense and turbulent cloud of rock particles and water almost has a life of its own, flowing along the ocean bottom over great distances. The turbidity current is often confined to underwater canyons that form deep scars in the seafloor, and it can be powerful enough to pluck bedrock from the sea floor, deepening and widening the canyon. Eventually, the cloud settles as a new layer on the seabed, forming a distinctive deposit called a turbidite.

A pile of turbidites many kilometers thick can easily build up over geological time— say, a few million years—producing a type of bedrock long recognized in many mountain belts, though its origin was unknown until Heezen and Ewing's pioneering work. The key message is of an unstable and shaky world subject to frequent and violent natural disasters. And this turns out to be the world in which the sedimentary layers of the ancient Fig Tree Group of the Barberton Greenstone Belt were laid down.

ANCIENT AVALANCHES

Perhaps the easiest place to reach in my study area in Eswatini is a long, high ridge called Ngwenya—Siswati for "crocodile"—at the southern end of the Malolotja Nature Reserve. You can drive most of the way on a winding tar-sealed road, turning off the main highway into South Africa not far from the international border. The border itself is a wire fence running along the western flanks of the Ngwenya Ridge. The tar-sealed road going up to Ngwenya exists because it once served as the access route to a major iron ore mine. The mine opened in the 1960s, but by the time I started working in the reserve it was closing because all the best ore had been extracted. The scale of the operation is apparent in the large open pit and rust-red spoil heaps that have reshaped the southern end of the ridge. There is also an abandoned railway line, built exclusively for the mine, that once linked Ngwenya with the port of Maputo in Mozambique—about 160 kilometers away as the crow flies.

The iron ore is in a sequence of Fig Tree shales, siltstones and sandstones which can be traced for tens of kilometers along the international border, extending from Ngwenya in the south all the way to the Komati River and beyond. The Fig Tree layers at Ngwenya have also been twisted around, bending back on themselves to form a giant fold. I will come back to this fold in the next chapter, and so for the moment, all you need to appreciate is that the same layers can appear in many different places. Over the years, a number of geologists have looked at the Fig Tree Group—in particular, Gary Byerly at Louisiana State University, Don Lowe at Stanford University, and their students and co-workers. What they found extends our understanding to wider parts of the Barberton Greenstone Belt, although Isabelle Paris and I sometimes disagreed with them on details of interpretation. Here, I will just focus on our own work because this is where I have first-hand knowledge and I think we can work out the essence of what happened at this stage in the geological story I want to tell.

Isabelle was studying the rocks just across the international border from me, in South Africa. By standing on the highest point of the Malolotja Nature Reserve, I could look directly into her field area. She had found essentially the same rocks as those at Ngwenya, recognizing that the iron-rich layers were the bottom part of what looked like a colossal pile of submarine debris avalanches and turbidites, with all the characteristic features of this type of sediment. When I visited her in her field area, she showed me what she had found. This alerted me to the same sequences on my side of the border in Swaziland. We decided that the iron-rich layers at Ngwenya were deposited in deep water, where only mud and silt were likely to settle out. The precipitation of the iron may have taken place near the sea surface, perhaps caused

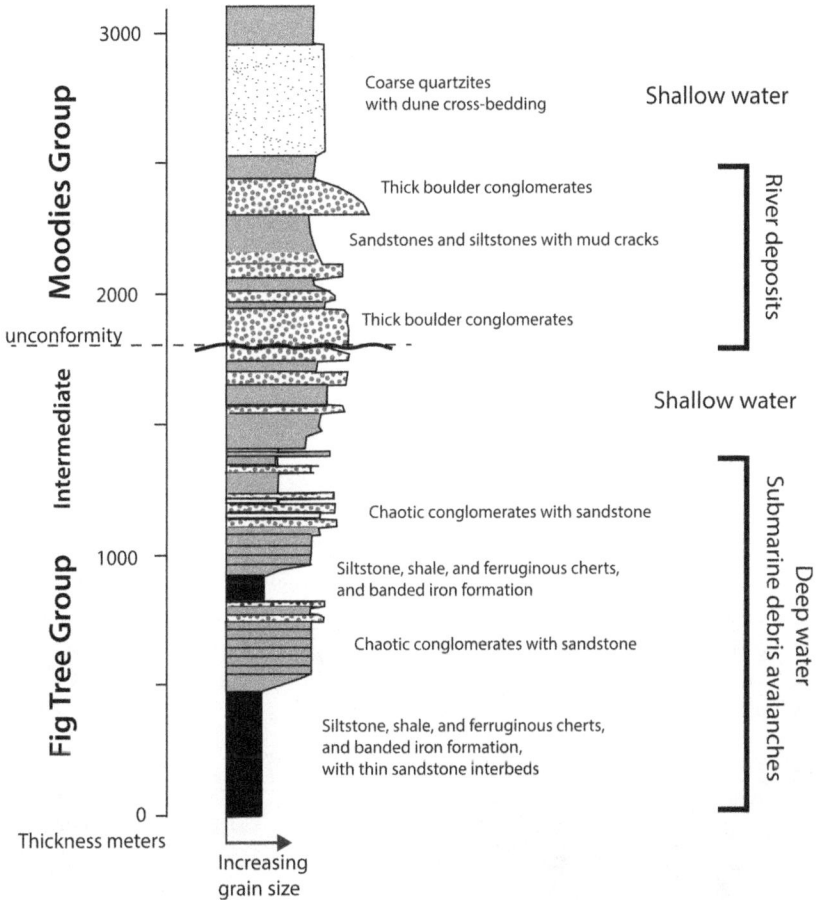

Sequence of rocks in
Moodies and Fig Tree Groups
at Ngwenya, Eswatini

Moodies Group

3000

Coarse quartzites
with dune cross-bedding

Shallow water

Thick boulder conglomerates

Sandstones and siltstones with mud cracks

River deposits

2000

Thick boulder conglomerates

unconformity

Shallow water

Intermediate

Chaotic conglomerates with sandstone

1000

Siltstone, shale, and ferruginous cherts,
and banded iron formation

Chaotic conglomerates with sandstone

Submarine debris avalanches

Deep water

Fig Tree Group

Siltstone, shale, and ferruginous cherts,
and banded iron formation,
with thin sandstone interbeds

0

Thickness meters

Increasing
grain size

FIGURE 4.6 The layers of Fig Tree and Moodies Group sedimentary rocks, exposed on the Ngwenya ridge at the southern end of my study area in the Malolotja Nature Reserve, reveal a profoundly unstable world marked by extreme tectonic violence. During Fig Tree times, earthquakes shook the ocean margin, triggering underwater debris avalanches that cascaded down into the deep ocean, piling up as turbidites and chaotic conglomerates on top of deep-water muddy sediments. I argue that the forces of plate tectonics in a subduction zone drove this activity, eventually lifting up the seabed until it emerged as land. Rivers then flowed across this new land, depositing the sandstones and conglomerates that now make up the Moodies Group. (Illustration credit: Simon Lamb)

by the metabolic activity of blooms of microbes exposed to sunlight—just as I described for the formation of the ferruginous cherts through the process of photoferrotrophy—perhaps during the summer months when the water was warm. The heavy iron particles would have then rained down through the water column and accumulated on the deep seafloor, where they were subsequently buried by turbidity currents that flowed down from a much shallower seabed. All this was a clue that these rocks had been laid down on the unstable edge of a deep ocean, just like the debris avalanche triggered by the 1929 Grand Banks earthquake. This was a very different place from those shallow waters and volcanic landscapes where stromatolites and mud pools could be found.

But there was even more dramatic evidence that a profound change was taking place in the stability of Earth's surface. Much higher up the stack of turbidites there were layers made up entirely of large blocks of

FIGURE 4.7 Looking into Eswatini from South Africa, showing the rugged skyline ridge that marks the international border and forms the northern part of my study area, in the Malolotja Nature Reserve. This ridge is underlain by thick, chaotic conglomerates of the Fig Tree Group, laid down in deep water. (Photo credit: Simon Lamb)

chert—many of them quite sharp sided—up to half a meter across and arranged higgledy-piggledy; sometimes slabs of chert were leaning against each other, a bit like a stack of loose roof tiles resting on their sides against a wall. Geologists call this type of rock a conglomerate—just like an industrial conglomerate with many international divisions, it too is a mixture of diverse components. In places, the conglomerates were chaotic piles, hundreds of meters thick, of jumbled-up blocks. But these were more like pods than layers, becoming thinner in both directions as you traced them through the landscape.

Isabelle and I identified one particularly distinctive and massive pod, up to half a kilometer thick, which reappeared in a number of places within our respective study areas. When we joined them all up, they seemed to form a great "snake" of conglomerate wriggling in a roughly north–south direction, oblivious to the international border between South Africa and Swaziland. We interpreted this as the debris created by the destruction of the underlying chert layers and volcanic rocks, trapped in a long submarine canyon full of rubble. Here, we differed from the American group studying the Fig Tree Group elsewhere in the greenstone belt, who considered these conglomerates to have been deposited by rivers on land. They had found other indicators of shallow water or dry land—such as evidence of mud that had dried out and cracked—farther to the north. But we could see no break between land and sea in the sequence we were looking at, and the same iron-rich siltstones and shales found above the conglomerates indicated that we were still in deep water. But what really convinced us was that the conglomerates had all the features expected for our submarine canyon, as worked out by geologists and oceanographers who have studied these types of rock—both ancient and modern—in many parts of the world.

But I agree that land and rivers cannot have been far away. The large, sharp- or angular-sided blocks indicated that the rubble had not been carried far in a water current—perhaps no more than, say, ten kilometers—as greater transport would have caused wear and tear, breaking up the blocks and smoothing their surfaces. We think that we are close to the site of the landslides that triggered the submarine debris avalanches and turbidity currents. The signs of catastrophic collapse were becoming clearer to us. But I only appreciated the full significance of all this when I had the

FIGURE 4.8 A typical conglomerate in the Fig Tree Group, forming layers in the great fold at Ngwenya, at the southern end of my study area in Eswatini. The conglomerate is made up of sharp-side blocks of chert, plucked from the Onverwacht Group by powerful underwater debris avalanches and currents. Later squeezing of the rock has changed the shapes of these blocks, so that they are flattened and stretched out. (Photo credit: Simon Lamb)

opportunity to observe firsthand the sort of place where such a collapse could happen today—not on the Grand Banks, but in the much more earthquake-prone region of New Zealand.

GIANT PLUM STONES IN A PUDDING

After I had completed my PhD at Cambridge University, I applied for a postgraduate fellowship at Victoria University of Wellington, New Zealand. Because I had spent the previous three years immersed in some of the oldest rocks on the planet, trying to reconstruct the ancient world they represented, I jumped at the chance to switch my focus to the other end of Earth's history to study some of the youngest. New Zealand straddles two tectonic plates. Here, the Pacific tectonic plate, which underlies much of the southwest Pacific, is sliding—or subducting, as geologists say—beneath the Australian Plate in a subduction zone. The Australian Plate comprises the continent of Australia, the bed of the Tasman Sea to the east, and much of western New Zealand.

One of my first tasks in New Zealand was to try and work out when the current phase of subduction in New Zealand started. I was fortunate to meet Harold Wellman, one of the country's most distinguished geologists. He had played a major role in discovering the many fault lines in New Zealand's North Island and South Island, and in understanding how they fit into the theory of plate tectonics. Harold was retired, but he continued to come into the geology department for morning tea, still startling its members with new ideas. When I told Harold about my project, he suggested I go and look at a peculiar geological formation in the northeastern part of South Island, not far from the small town of Kaikoura. This formation is known as the Great Marlborough Conglomerate, partly because it lies in the Marlborough district, but more importantly because it is truly "great," as you will see!

Coming from a research project in southern Africa, I was intrigued to find out that the Great Marlborough Conglomerate had been extensively studied by a geologist named Lester King who was also the author of what is now considered to be the standard work on the geological features of the South African landscape. King was a New Zealander, but just before World War II he had taken a job at the University of Natal in South Africa, where he spent the rest of his career. His work on the Great Marlborough Conglomerate was presented in a scientific paper published in 1937. Despite avidly reading this several times over, I was still taken by surprise when I went to have a look for myself. My timing was not the best, because it was in the depths of a New Zealand winter—some streams had a layer of ice in the early morning, and the days were short. The region consists of rugged grassy hills, farmed for sheep and cattle, with a backdrop of the snow-covered peaks of the Kaikoura Ranges, rising to 2,885 meters above sea level. The Great Marlborough Conglomerate is the collective name for the strange bedrock of this country, made up mainly of a chaotic mixture of mudstones, turbidites and conglomerates, just like the sedimentary rocks

Great Marlborough Conglomerate, New Zealand

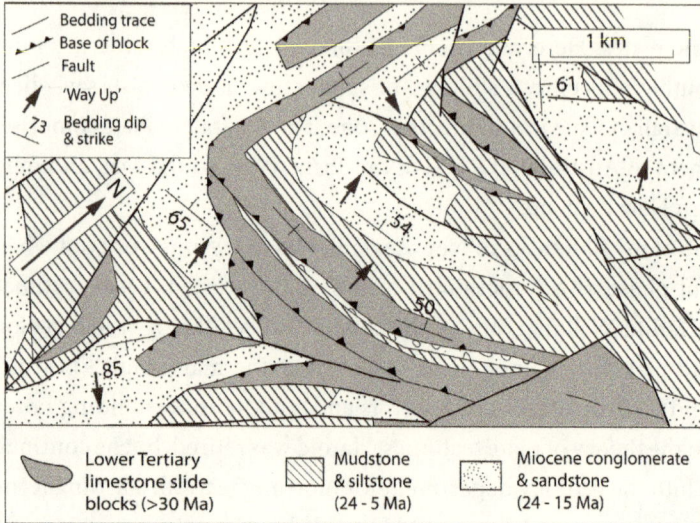

Bedding trace
Base of block
Fault
'Way Up'
Bedding dip & strike

1 km

Lower Tertiary limestone slide blocks (>30 Ma)

Mudstone & siltstone (24 - 5 Ma)

Miocene conglomerate & sandstone (24 - 15 Ma)

Central Barberton Greenstone Belt

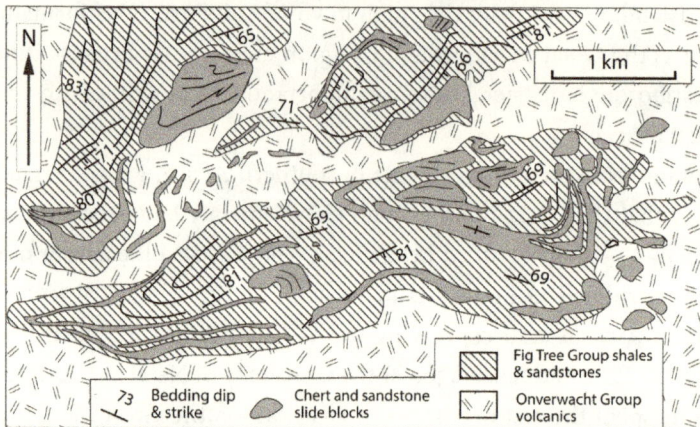

1 km

Bedding dip & strike

Chert and sandstone slide blocks

Fig Tree Group shales & sandstones

Onverwacht Group volcanics

FIGURE 4.9 My mapping of the rocks in New Zealand's Great Marlborough Conglomerate showed that they are the remains of giant submarine landslides and debris avalanches that slid down the steep edge of the continent into the ocean trench, in a subduction zone, about 10 to 20 million years ago. This map is remarkably similar to the one made by my colleague Cornel de Ronde in the central part of the Barberton Greenstone Belt, within the Fig Tree Group (see map of the Barberton Greenstone Belt in chapter 2 for the location of Cornel's map). He found huge blocks of chert and sandstone, hundreds of meters to kilometers across, often arranged topsy-turvy. All this suggests that these too are the remains of giant submarine landslides and debris avalanches on the edges of an ocean trench on early Earth. (Illustration credit: Simon Lamb)

in the Fig Tree Group back in southern Africa. These are visible in the banks of the many creeks, but there are also prominent hills or ridges of white limestone that stand proud in the landscape. It is the presence of this limestone that makes the Great Marlborough Conglomerate truly great. Let me explain.

Limestone can be traced over much of New Zealand. It was laid down on the margins of a stable landmass close to sea level, going back to the end of the Cretaceous Period, during the time of the dinosaurs. In fact, the traces of the asteroid impact, which may have been instrumental in the extinction of the dinosaurs and many other forms of life 66 million years ago, is preserved within the limestone as a thin layer enriched in the rare element iridium. Subsequently, more limestone was deposited in the surrounding shallow seas. However, around 20 million years ago, at the beginning of the Miocene Epoch, everything seemed to change drastically. This was when the Great Marlborough Conglomerate burst upon the geological scene.

King recognized that many of the rocky ridges in the landscape are actually isolated blocks of limestone *surrounded* by the other sedimentary rocks, like giant plum stones in a pudding. This way, the bedrock truly lives up to its collective name as the Great Marlborough Conglomerate. It is relatively easy to accept this for some of the smaller limestone blocks—although they are still the size of a house—because mudstone, sandstone and pebbly conglomerates can be seen in the banks of all the neighboring streams. Deep-sea fossils in the mudstones show that these were laid down offshore, in deep water. The pebbles may have originally been carried by rivers all the way to the shelf edge when sea level was lower relative to the land. But when I tried to make my own detailed geological map, I realized that some of the limestone blocks must extend for nearly ten kilometers—about the size of a small island in, say, the Scottish Hebrides or the Caribbean. Sometimes I found it difficult to tell if the limestone was an intact part of the underlying bedrock, or a block in the conglomerate. It looked to me as if great chunks of the older bedrock had literally started to fall apart, sliding for kilometers on a lubricated cushion of mud down the steep edge of the continent into the deep ocean. There were examples where one limestone layer had clearly slid down on top of another, with patches of mud and turbidites caught up in between.

I think we are witnessing, in the Great Marlborough Conglomerate, the birth of a subduction zone. I should clarify that this does not mean there was no plate tectonics prior to this—we know that plate tectonics was going on long before—but rather that this was the birth of a new phase of subduction at this particular location on Earth. And it was a difficult birth. The Pacific Plate must have started to break apart, with one side sinking back into the underlying mantle, pulling down the seafloor with it to form a deep ocean trench. This would have triggered earthquakes that violently shook the steep edges of the adjacent land. The Great Marlborough Conglomerate is the catastrophic consequence of this, created during numerous earthquakes by gigantic landslides and debris avalanches that continued to slide into the newly formed trench.

The region is still very unstable, and only a few tens of kilometers offshore, near Kaikoura, the seafloor drops away in the Kaikoura Canyon into an ocean trench about four kilometers below sea level. This trench is part of a series of subduction zones that ring the Pacific Ocean, where the Pacific Plate slides beneath its margins, and it is one of the most earthquake prone places on the planet. In November 2016, a magnitude 7.8 earthquake near Kaikoura set off underwater landslides and debris avalanches that travelled over 680 kilometers down the underwater Kaikoura Canyon. It is possible to work out what happened by comparing detailed surveys of the seabed before and after the earthquake. About 850 million tons of rock shifted into the deep ocean in the few hours following the earthquake, leaving a gash approximately 50 meters deep in the head of the canyon. And the huge scars along the edge of the ocean trench show that previous earthquakes have long been destabilizing this region, gradually nudging sliding blocks ever farther along their journey into the ocean depths.

THAT SINKING FEELING

I am struck by the similarities between the geologically sudden destruction of the shallow seas and bedrock around New Zealand, with the creation of a steep continental margin, and the events in the Barberton Greenstone Belt when the Fig Tree sedimentary rocks were deposited. This was brought

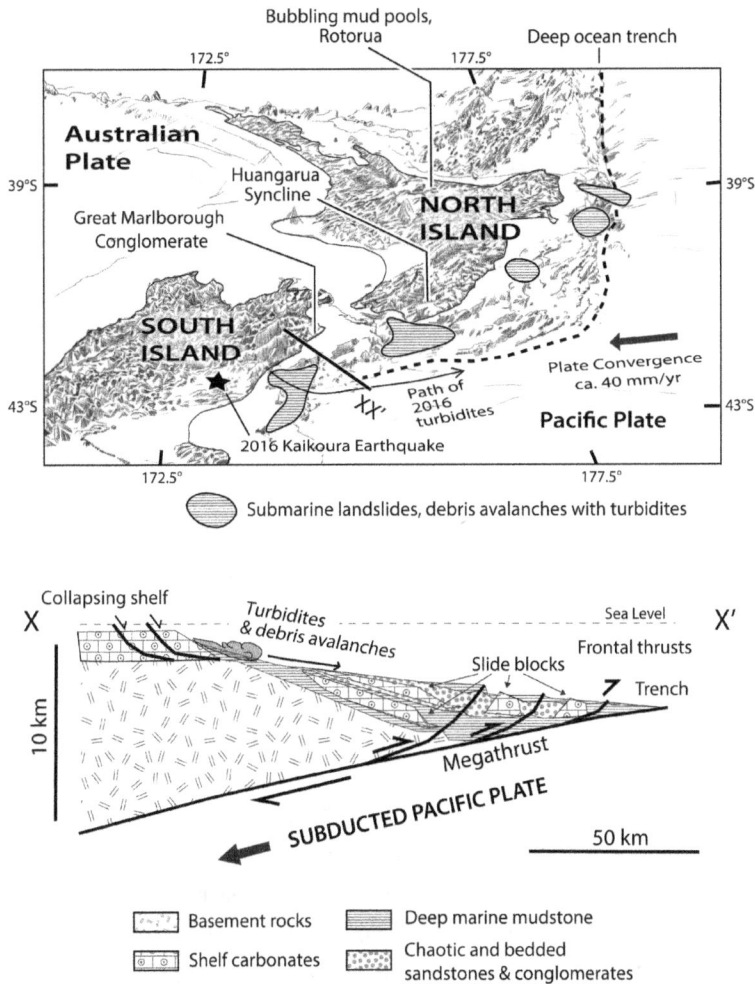

FIGURE 4.10 New Zealand may be our best analog for the ancient world during Fig Tree times, as preserved in the Barberton Greenstone Belt. Today, submarine landslides and debris avalanches are forming just offshore, where the shallow shelf is collapsing along the steep edge of the ocean trench and the Pacific Plate is sliding along a megathrust, underneath New Zealand, back into the mantle— uplifted examples of these underwater landslides and debris avalanches are preserved onshore in the Great Marlborough Conglomerate. They are triggered by earthquakes, such as the magnitude 7.8 Kaikoura Earthquake in 2016. Every few hundred years, the shaking from a large earthquake sets off debris avalanches and associated turbidites, or nudges giant blocks in landslides a little farther along their journey into the deep trench; thousands of earthquakes were most likely required to form both the geologically young Great Marlborough Conglomerate and the very ancient Fig Tree Group. New Zealand affords other opportunities to see conditions on early Earth—for example, the bubbling mud pools near Rotorua, North Island, and the Huangarua Syncline farther south (described in chapter 5). (Illustration credit: Simon Lamb and Clara Maxwell Lamb)

home to me recently when I helped prepare a new, detailed (1:14,100 scale) geologic map of the central part of the Barberton Greenstone Belt for publication. The map was made by Cornel de Ronde during his PhD and post-doctoral research, over a period of nearly eight years. I think it reveals a previously unrecognized feature of the rocks: the Fig Tree Group largely consists of blocks of chert and ferruginous shale, ranging from hundreds of meters to several kilometers across, lying topsy-turvy—either surrounded by shale or sitting on komatiite lavas. The sides of these slabs or blocks are sharp, and the internal layering is sometimes strongly rotated relative to the enclosing shales and sandstones. Importantly, layers or blocks of sedimentary rocks that were deposited in shallow water, or on land, are immediately above or below those that formed in much deeper water.

This has all the appearance of the jumbled rocks in the Great Marlborough Conglomerate in New Zealand. In fact, if you were to take my geological map of the Great Marlborough Conglomerate and replace the limestone blocks with those of chert, you would have something that looks remarkably like Cornel's geological map, both in scale and in overall pattern. Just as in New Zealand, it looks as though whole portions of the underlying bedrock, several kilometers long, have slid into the deepening seafloor of young Earth, and now lie out of order on top of each other. This may be the origin of the famous pylon nappe of chert that Maarten showed to me and Isabelle on our first introduction to the greenstone belt, described in chapter 2. What we were looking at through our binoculars, perched on an adjacent hill, may be direct evidence for this sliding—revealed in the way the large slab of chert had bunched up and folded back on itself like a caterpillar track. This image remains so vivid for me that I can almost feel it moving today.

Let me summarize where I think we have reached in the story of the young planet, a sequence of events that we have deciphered from the rocks in the Fig Tree Group. We are on the edges of dry land—the Kaikoura coastline and offshore seabed of New Zealand is a clue to the proximity of this land to the deep ocean, although the nature of the bedrock is very different. On early Earth, the land was underlain by a thick pile of lavas and cherts that had either built up on the seabed, eventually reaching above sea level, or that had been pushed up by forces within Earth. Rivers from the interior carried their load of mud, silt, sand, and gravel to the coast, from where the sediment was eventually taken by currents out to the deep sea. In very shallow water, colonies of photosynthesizing microbes thrived, making their characteristic stromatolite mounds. Signs of intense volcanic activity were everywhere: bubbling mud pools, geysers, still-warm lava flows. Occasional meteorite impacts and earthquakes shook the land, opening gaping chasms into which volcanic ash fell.

Don Lowe at Stanford University has suggested that what happened next was set in motion by one of those violent meteorite impacts. I prefer to think that earthbound forces were at work, deep inside the planet, driven by the push and pull of gravity and the churning of convection in the mantle. The ocean floor began to buckle under this pressure, and was forced down. The thick oceanic crust now experienced the higher temperatures and pressures deeper in the underlying mantle. These conditions would eventually transform the minerals in the rock into new and more compressed forms, creating a heavy rock called eclogite that has an even greater tendency to sink. In this way, I think, a new subduction zone was

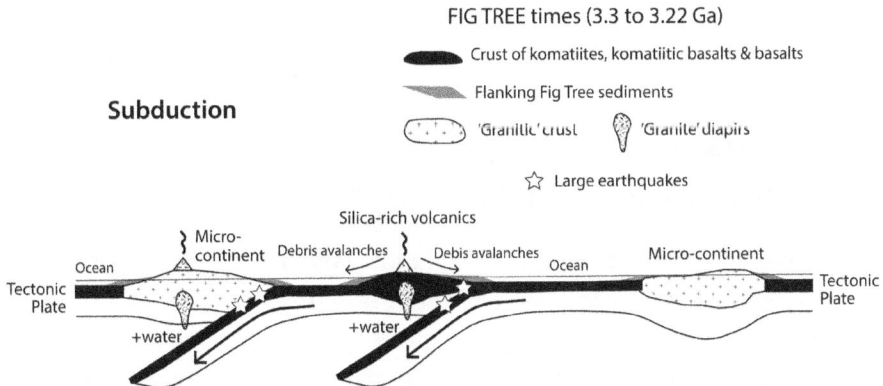

FIGURE 4.11 Hypothetical diagram illustrating how plate tectonics might have worked during Fig Tree times. Subduction of tectonic plates would have triggered large earthquakes, violently shaking the edge of a volcanic arc or ocean trench and setting off underwater landslides and debris avalanches. Subduction was also bringing together two small continents that would eventually collide during Moodies times, raising a mountain range. (Illustration credit: Simon Lamb)

born. And as with the newly created subduction zone in New Zealand—at the time of the Great Marlborough Conglomerate—the sinking plate pulled down the adjacent parts of Earth's surface, tearing at the foundations of the dry land.

All this activity triggered even larger earthquakes that profoundly destabilized the remaining coastal regions. Huge landslides and debris avalanches were set in motion, filling up canyons in the seabed with turbidites and chaotic conglomerate. Blocks of the shallow-water bedrock started sliding down into the deep ocean, where they piled up on the sea floor. At the same time as parts of early Earth's surface were dramatically falling, other parts were beginning to rise, creating mountain ranges with many similarities to the great ranges we see today. And the nature of Earth's crust was also starting to change in profound ways, forming the early continents. I think the rocks in the Makhonjwa Mountains are telling us what happened when parts of early Earth's surface got caught up in the viselike grip between converging tectonic plates, either in a subduction zone or when continents collided. This is the story I will tell in the next chapter, now that the scene has been set. But I will leave you here with some more tales of my own fieldwork in Swaziland.

- § -

CAMPING OUT AND LONG GRASS

My time roaming the Malolotja Nature Reserve, mapping the rocks, was a very solitary experience. For long stretches of time—up to two weeks or more—I was entirely on my own. Yet I never felt lonely. My mind was completely taken up with geology, and I was immersed in a world of rocks. Getting to these rocks required a detailed knowledge of the landscape, and I must have walked many miles every day, ascending and descending up to a thousand meters as I explored new parts of the reserve. A constant theme of these long days in the field was the effort of pushing through tall grass and the distinctive, pungent aroma of aromatic plants—like curry bush and wild rosemary—given off whenever I brushed against them. Even today,

FIGURE 4.12 A rare view of myself striding through the tall grass in the Malolotja Valley, Eswatini, wearing a sun hat and carrying a notebook, field maps, and a pack laden with rock samples. (Photo courtesy of Simon Lamb)

regardless of where I am in the world, the scent of these plants immediately transports me back to the hot, shimmering days in my old field area.

The big decision for me was where to base myself in the field. I rotated between several camps, most of them located at the ends of access tracks deep in the reserve. The tracks were no more than a parallel set of wheel marks, and sometimes the track could be lost altogether in the long grass. During my first full field season, I purchased an extremely old and unreliable short-wheelbase Land Rover in Johannesburg. This in itself was no easy task, as Land Rovers were in high demand in South Africa at the time—mainly among farmers and the military. It was soon clear that the one I had managed to buy was a lemon; it consumed almost as much oil as petrol. I used to joke that whenever I pulled into a gas station, I'd say, "Fill it up with oil—and could you check the gas too?"

In the very early days, I set up a tent for my camp. But one night I was spooked by a herd of wildebeest that were grazing nearby. Several times during the day they had charged at me in unison, and I began to feel unsafe near them. That night I had a nightmare in which the wildebeest stampeded my camp, trampling my tent with me inside. I remember waking up screaming, convinced that a wildebeest had just put his hoof on my stomach—in fact, my pack had fallen on me. I abandoned the tent and tried to sleep in the Land Rover instead, stretched uncomfortably across the front bench seat. From then on, I lived out of the Land Rover. But at the end of the field season, when I returned to Johannesburg to catch my flight back home, I got rid of the oil-guzzling lemon. I bought Maarten's old long-wheelbase Land Rover instead, which was in much better condition and had more room in the back. Maarten was leaving South Africa to spend a year in Canada, and so he was pleased to be able to dispose of it so easily. The new Land Rover offered considerable advantages over the old one. I could now sleep in the back of it, comfortably stretched out on a camp bed. I arranged all the cooking equipment on a side bench. I used to look forward to returning to my "home" after a hard and hot day laboring up and down steep hillsides with a backpack full of rock samples.

I found I could get very good reception for the English radio service of the South African Broadcasting Corporation. I tuned in regularly to several programs, including the news, a book serialization, and comedy sketches. The news rarely looked beyond national borders, and it was frustrating to only hear brief comments on important international events. Once, as I was cooking dinner in the heart of the Malolotja Nature Reserve, I listened to a lead story on the six o'clock news about an old lady in Bloemfontein who had been blown over in strong winds. I still think about that old lady. But it was at the time of the Falklands War, in 1982, and I was desperate to know what was happening. There were only the briefest statements about it—I suspect that the South African government did not want to appear to be supporting Britain over Argentina. In any case, the domestic news was generally rather depressing, usually about yet another round of failed talks on resolving the problem of Apartheid. At the time, any political resolution seemed hopeless. It would have been hard to believe that ten years later South Africa would finally wake up from its self-imposed Apartheid nightmare. I will write more about what it was like to operate in this cruel Apartheid world at the end of a later chapter.

I remember realizing after a couple of weeks camping out on my own that I had not uttered a single word in that whole time. I suddenly wondered if I had lost the power of speech. With some trepidation, I spoke out loud and was relieved to hear that I could still put a sentence together. I used to take delight in "first-time-ever" events for my camp. I had a tape player and a large collection of music tapes. I would play favorite pieces of classical music to the landscape at full volume—I was particularly keen on Mahler and Sibelius—amusing myself with the thought that this must be the first time in the history of the world that these sounds had been heard in this place, right in the middle of the Malolotja Nature Reserve, on the edges of the Makhonjwa Mountains, among some of the oldest rocks on Earth. Although not something for the Guinness Book of Records, it was, for me, a demonstration that it is still possible to do something that nobody else has ever done before.

In my second field season, the grass was particularly long, and the warden was organizing controlled burns. At night I could see red and orange glowing lines of burning grass in the distance. It seemed to me to be a sort of vision of hell. I was camped in the wonderfully named Mhlamgamphepha Valley (pronounced Um-lam-gam-PEAR-pa), and I remember wondering one night whether it was sensible to remain where I was—I would be in a dangerous situation if the grass around my camp caught fire. I suddenly woke up feeling afraid. I had an image of the fire getting so close to the Land Rover that it ignited the gas tank, which then exploded, destroying the vehicle. I made the decision to pack up camp and leave right away. As I maneuvered the Land Rover in the dark, trying to make out the line of the track, I could see flames all around me on the hilltops. I must have escaped just in time, because when I returned the next day, my old camp site was a smoldering desert of charred ground. There was one advantage to these burns: they revealed the bedrock and I could easily trace out the lines of strata in the burnt hillsides.

The most eerie camp I had was right in the heart of the Malolotja Valley, which formed a sort of bowl in the middle of the reserve. It felt haunted by the ghosts of villagers who used to live there. All around I could see the remains of their houses and fields, and the many paths connecting them. On hilltops overlooking the settlements, there were piles of stones marked with strange symbols. I was told that these were burial grounds where the

FIGURE 4.13 The Malolotja River winds through its deep valley in the Malolotja Nature Reserve, Eswatini, on a typical day with low morning clouds. The only access is on foot or by following faint wheel tracks through the tall grass and straight down the steep valley side. The white circle marks my Land Rover—just a white speck—parked at my usual camping spot. (Photo courtesy of Simon Lamb)

locals believed there was powerful magic, and which they would visit from time to time—I steered well clear of them. I found a convenient place in the valley for my camp, away from the old settlements and right next to the Malolotja River, although I had to crash through the wide, thickly vegetated river flats each morning to reach the rocks. In my first year I stayed on until late in the season—into early spring in the southern hemisphere, when the days start to heat up. I began to see huge snakes curled up on rocks, just emerging from their winter sleep. I could also hear the snakes jumping out of my way as I approached. I was telling the warden about these experiences when he remarked that the Malolotja Valley has one of the densest populations of black mamba snakes in southern Africa. You should know that the black mamba has a venom that is fatal unless you can get the antidote in less than twenty minutes. This had been my favorite camp, despite its eeriness, but thereafter I had very different feelings about it.

Small insect-like creatures were probably the biggest threat. Once, I was staying in one of the rangers' huts in a remote part of the reserve. This was just a concrete blockhouse with a corrugated iron roof, located right next to the Komati River where it enters Swaziland from South Africa. I had spent the day on the hills above the hut, and I came down hot and thirsty. After dinner, as I lay down on my camp bed, I became aware of a lump near my waist. It seemed to be about the size of a fully ripe grape, and it was attached to me. This did not make any sense, so I stood up, pulled my shirt up, and shined a flashlight on myself. The lump was bright red, just like the grape I had imagined. But it wasn't a grape—it was a tick that had gorged itself on my blood. I knew that if I pulled it off, it would probably leave its jaws in my skin, and the bite could become infected. There was also the danger of contracting tick-bite fever or meningitis. In the end, I used a candle to burn it off. Afterwards, when I viciously squeezed it, blood squirted out. Fortunately, that was the end of the matter, and the bite soon healed up.

The large animals in the reserve were always the subject of most interest for anybody I spoke to. I had been talking to the warden about how I had only seen wildebeest, springbok (a type of antelope), zebra, baboons, and the occasional warthog. There were supposed to be giant monitor lizards along the banks of the Komati River, and a very rare bald ibis nested in rocky crags where the Malolotja River cascaded down a waterfall into its deep valley. The warden told me that he wanted to increase the variety of mammals, and he had plans to introduce giraffes, elephants, and big cats. After learning about the black mambas, I can't say I was thrilled about the prospect of having big cats on the loose, too. But I was assured they would never come near, based on the assumption that they would be more afraid of me than I was of them. The warden then told me with great satisfaction that he had already been promised a leopard from a game reserve in the Lowveld of Swaziland.

The day soon came when the leopard was to be set free. It was now growling in a wooden crate with a sliding door at one end, sitting in the back of a Land Rover. I accompanied the warden as he took it deep into the reserve, in the heart of the Malolotja Valley and right next to my old camping spot. The track is very steep, and the crate got knocked about in the back of the vehicle, no doubt further upsetting the animal. The warden

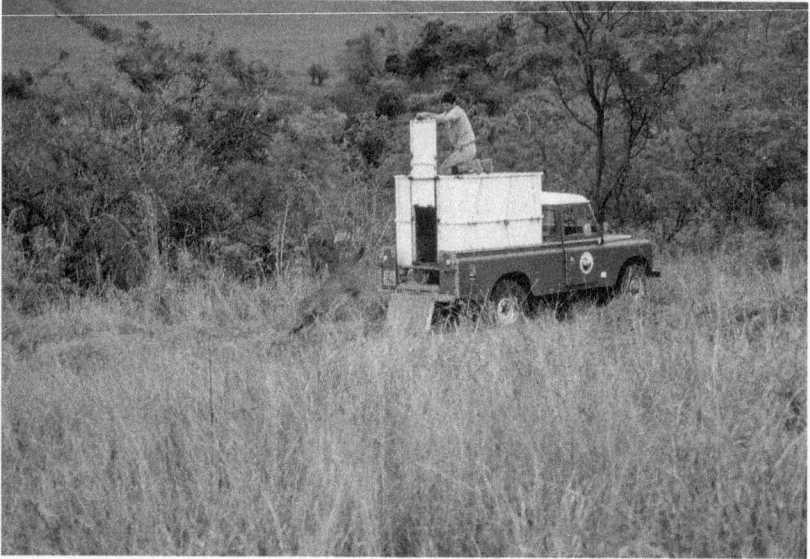

FIGURE 4.14 Disappearing act. A leopard is no more than a blur as it leaps from its cage upon release into the Malolotja Nature Reserve, Eswatini. It was never seen again. (Photo credit: Simon Lamb)

climbed on top of the crate and slowly slid open the door so there was now nothing between the leopard and freedom. I noticed that the rangers were not taking any chances, and they had their rifles ready. The leopard took a few steps forward, so that just its head poked out. It looked from side to side, then stood very still for what seemed like several minutes. We all had our cameras ready to take the defining picture of this magnificent animal, once it had fully emerged from the cage. Then suddenly, without any warning, the leopard gave out an enormous roar and was gone—my picture is just a blur. And that was the last anybody ever saw of it.

5 | Shifting Landscapes

The first continents: rising granites and raising mountains

Our world is a peculiar place. Only one-third is dry land—the rest is water. If we could drain the water, we would find that the dry land mainly forms isolated regions which rise many kilometers above the ocean floor. These, of course, are the continents. But have you ever wondered *why* our world is like this? Why not a single, jumbled landmass with many land-locked seas, or a planet completely underwater? We can summarize our questions in a simple way: what determines the level of Earth's surface? Answering this is the key to understanding what we will find next in our time travels, as we move away from the early oceans and the shorelines at their edges, and venture inland to explore the young continents. It was astronomers rather than geologists who first began to think about these aspects of Earth. And their conclusions have allowed us to track the rise of some of Earth's first mountain ranges, guided by the rocks in the Barberton Greenstone Belt, exposed in the Makhonjwa Mountains. Our story begins in Cape Town, at the southern tip of the African continent.

ASTRONOMICAL BRAINWAVES

Cape Town is renowned for its clear skies, swept clean by the winds that circle Antarctica, and in the nineteenth century it became a frequent destination for astronomers and other scientists. Most famously, the astronomer John Herschel spent time here in the 1830s, observing Halley's Comet and the distant stars. (You may be more familiar with his father, William Herschel, who discovered the planet Uranus.) John Herschel remarked that his stay in Cape Town was the happiest time of his life: free from the pressures of his London commitments, he could think more broadly about science. He used this freedom to explore ideas about the nature of Earth's surface, drawing inspiration from the spectacular and rugged landscapes behind Cape Town—including the high plateau farther inland known to Herschel as the table-lands (not to be confused with the famous landmark Table Mountain). He also maintained written correspondence about his ideas with other leading scientists of the day.

Herschel's interest in Earth and other planets reached far beyond astronomy. He was aware that the shape of Earth is virtually identical to a giant spinning liquid ball, pulled together by the force of gravity and distorted by the forces of rotation. This has made the planet bulge around the equator and become slightly flattened at the poles. The detailed mathematical theory had been worked out nearly a century earlier, in 1743, by the French mathematician Alexis Claude Clairaut, building on Isaac Newton's pioneering work on gravity. Herschel believed that Earth was molten inside, whereas nowadays, geologists recognize that even solid rock can flow over geological timescales. What Herschel correctly grasped, however, was that the interior of Earth behaves like a fluid, and this would need to be taken into account when trying to explain many other aspects of its surface.

While still in Cape Town, Herschel wrote to Charles Lyell—the leading geologist of the day—speculating "on the formation of new continents" and expressing a remarkable idea about these continents "supposing the whole to float on a sea of lava" like some sort of rocky boat. These ideas set the scene for a breakthrough in our understanding of Earth, although it was his colleague George Airy—then the British Astronomer Royal—who published them as a full-fledged scientific theory. I think it is no coincidence that George Airy was thinking along similar lines to Herschel, given that we know Herschel was communicating with him from Cape Town—and Airy was clearly trying to explain the origin of table-lands like those north of the city.

Both Herschel and Airy were building on the work of Archimedes, who, according to legend, was the first to figure out why objects float. Taking a ship as an example, the basic idea is that the hull sits in water up to the level where the weight of displaced water balances the ship's weight. A simpler way to say this is in terms of the density of matter, which measures how much weight—or more accurately, mass—is concentrated in a given volume. If you remember your high school physics, density is defined as mass divided by volume. Therefore, a ship floats if its average density is less than that of water. Two things make a ship rise higher above the waterline: being lighter or being scaled up in size. Thus, given similar ship densities, larger ships displace more water than smaller ones, but their decks are also higher. It was not until 1855 that Airy was ready to apply this concept in what is arguably one of the most important scientific papers in geology. He had now laid out the basic under-standing of what controls the elevation of the planet's surface. Following in Herschel's footsteps, he had made sense of the existence of continents and oceans on a planet that is fluid-like on the inside.

Airy started out by considering another problem: in the first half of the nineteenth century nobody was sure of Earth's mass. In effect, Airy wanted to weigh our planet with celestial scales. He knew the volume of Earth from its basic dimensions; if he could work out its average density, then he could calculate its mass. He came up with an ingenious method that relied on measuring changes in the force of gravity down a deep mine shaft, which he showed mathematically could, in principle, reveal the *ratio* of the densi-ties of the planet's outer and inner layers. For the former, he could measure density directly from samples of typical rock in the mine, and then with his newly determined ratio he could then calculate the density deep inside. The measurements were very difficult to make, but they showed that Earth's overall density was about six and a half times the density of water—the modern estimate is closer to five and a half times, so Airy's estimate was reasonably accurate, considering the crudeness of his instruments. The mass of Earth works out to be a staggering six billion trillion metric tons.

The lesson from all this, central to the argument in Airy's 1855 paper, was that the deep interior of Earth was denser than the outer layers. Airy also had access to some new data from measurements of the pull of gravity taken in northern India, near Mount Everest. Putting all this together, he had hard evidence that the outer, less-dense part of Earth, which he called the crust, must be floating on the underlying and more dense "fluid" interior—like Herschel before him, Airy actually wrote "lava." This implies that slabs of floating crust with different thicknesses will behave rather like floating ships of different sizes, with thicker slabs displacing more mantle and rising higher in much the same way that bigger ships displace more water and tower higher above the sea surface than smaller ones. In fact, Airy's analogy for the crust was that of logs floating on water. As he noted, if we notice "one log whose upper surface floats much higher than the upper surfaces of the others, we are certain that its lower surface lies deeper in the water than the lower surfaces of the others."

An iceberg is a good analogy too: like logs of wood, ice is less dense than seawater, and this is why icebergs float in the sea. And also like wood, the difference in density between ice and seawater is not very great, so only a relatively small part of the iceberg sticks up above the sea surface—about 10 percent of its total volume, to be precise. This had fatal consequences for

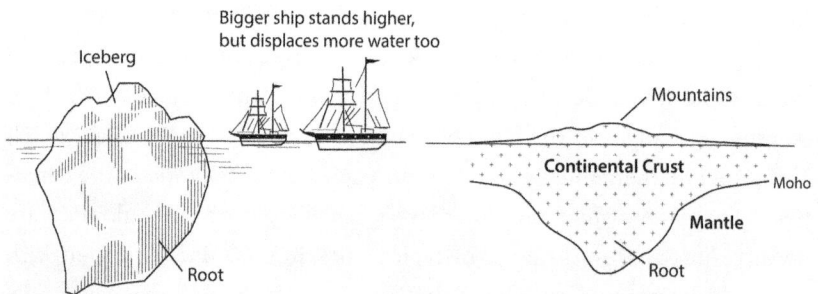

FIGURE 5.1 One of the great geological breakthroughs of the nineteenth century was the realization that Earth's crust is like an iceberg, or a ship at sea, floating on the denser, fluid-like mantle underneath. This means that for mountains to rise higher, they must have roots that extend deeper, resulting in a thicker crust compared to that beneath lowlands or oceans. (Illustration credit: Simon Lamb)

the *Titanic* in 1912, as the large submerged portion of a seemingly insignif-icant iceberg ripped a massive hole below the waterline in the ship's hull when it sailed too close. In effect, Airy was proposing that Earth's crust also has deep, iceberg-like roots—so if the surface rises higher, the roots extend deeper into the underlying fluid interior, resulting in a thicker crust.

Here was a fundamental explanation for continents and oceans. They were regions of thick and thin crust respectively. Yet when Airy first pro-posed his theory of a floating crust, nobody knew for sure whether Earth even had an outer layer in the way he was imagining—and they certainly had no idea how thick it was. However, Airy's thinking has stood the test of time. Geophysicists have now extensively probed the rocks deep below the continents and oceans with vibrations—called seismic waves—from either human-made explosions or earthquakes. By measuring the time taken for these seismic waves to travel through Earth it has been possible to work out their speed. Measurements on rocks in the laboratory show that different speeds are associated with different rock types and densities. This work has demonstrated that there is indeed a distinct outer layer to Earth where seis-mic waves are slower and the rocks have a lower density—this is the crust, overlying the mantle, and its base is called the Moho, after a Croatian seis-mologist Andrija Mohorovičić who first discovered it in 1909. The average depth of the Moho in the continents is about 40 kilometers. Beneath high mountain ranges it is much deeper, reaching depths up to about 80 kilome-ters, whereas it is only about 7 kilometers below the deep seafloor. And the crust is about 14 percent less dense than the underlying mantle.

The iceberg analogy is now obvious: the base of the crust is a greatly exaggerated mirror image of Earth's surface, like some distorted upside-down picture of the landscape reflected in an alpine lake, exactly as would be predicted if the crust was floating on the mantle.

JUST APPLY FORCE

It is clear that the existence of dry land, and mountain ranges that rise even farther, requires some mechanism for producing thick crust. And the scale of this thickening must be substantial, involving many-fold increases: typical continental crust is five to six times thicker than that beneath the oceans, and the crust beneath mountain ranges is twice as thick again. There is, however, one very simple way that this can be achieved—in fact, it can occur in a geological blink of the eye, over just a few million years or less.

The best way to visualize how this might happen is to think of a piece of uncooked pastry. You start with a lump of flour and butter. To make the thin layers of pastry you press down on it with a roller, and the lump flattens out. A thick lump, about as wide as your hand, can be pushed down into a thin layer that is as wide as the top of your kitchen table. Imagine running this process in reverse, taking a thin layer of pastry and lumping it back into a thick mound. In either case—thinning or thickening—you are trading height for width, one way or another, and the total amount of pastry has not changed. The same thing goes on inside Earth when the huge forces of plate tectonics push or pull on the crust.

Finally, I am ready to return to the rocks I was studying in the Makhonjwa Mountains of Swaziland and South Africa, which form the Barberton Greenstone Belt. At various points in my narrative I have mentioned

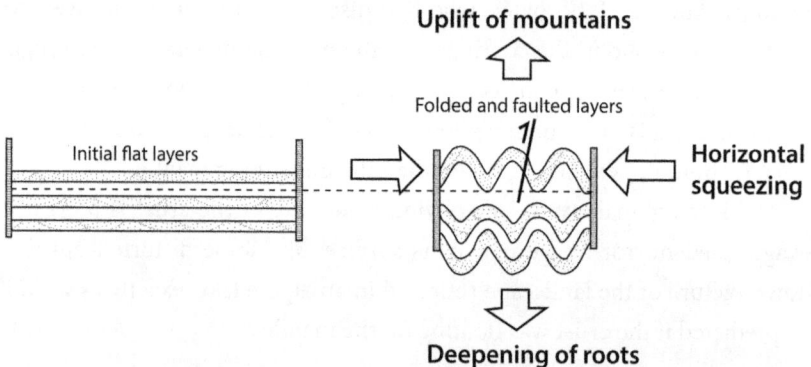

FIGURE 5.2 Swapping width for height. A geologically rapid way to thicken the crust is by squeezing it horizontally, creating folds and faults in the rock layers and pushing up mountains, but pushing down their supporting roots. This typically occurs between converging tectonic plates, either in a subduction zone or during continental collision.

the giant corrugations, or folds, in the rock layers—structures that run through the landscape, turning up the layers on end. In fact, the typical orientation of the rock layers in the greenstone belt is close to vertical. All of this clearly shows that the crust here has been squeezed together. An analogy might be squeezing a stack of flexible rulers by pushing their ends together until they buckle. If you do this too much, the rulers snap, rather like an earthquake when the rocks rupture along a fault. Inside Earth, you end up with folded layers with breaks or faults cutting across them, displacing the layers.

The fact that the layers in the Barberton Greenstone Belt are mostly vertical is a sure sign that this squeezing has been about as extreme as it can get. Squeezing like this is usually found above an even bigger fault, buried deep in the crust, such as the megathrust that separates the two tectonic plates in a subduction zone or in the region where continents collide. Here, because the overall volume of the crust does not change—or at least not by much—the horizontal squeezing results in a vertical thickening as width is traded for height, in much the same way as I described for bunching up a flattened layer of pastry. Counterintuitively, perhaps, this thickening of the crust causes most of it to be pushed downward, making those iceberg-like roots. But it is these deep roots that form the foundation for an uplifting mountain range at the surface. Roughly speaking, for every seven kilometers the roots extend downwards, the mountains rise up about one kilometer.

We can, then, interpret the folding and faulting of the rock layers in the Barberton Greenstone Belt as evidence for thickening of the crust at some point in the history of these rocks. The iceberg principle of a floating crust implies that this thickening would have led both to the emergence of new tracts of dry land as rocks rose above sea level and to mountain building as the land rose up even higher. But when did this happen? In particular, can we pin it down to a time when rocks in the Barberton Greenstone Belt were still being laid down, giving us direct evidence for dramatic uplift of early Earth's surface over 3.2 billion years ago? The answer is a resounding "yes." We can actually catch the rocks in the act of being squeezed and uplifted, creating a new mountain range.

Geological cross-sections through Barberton Greenstone Belt

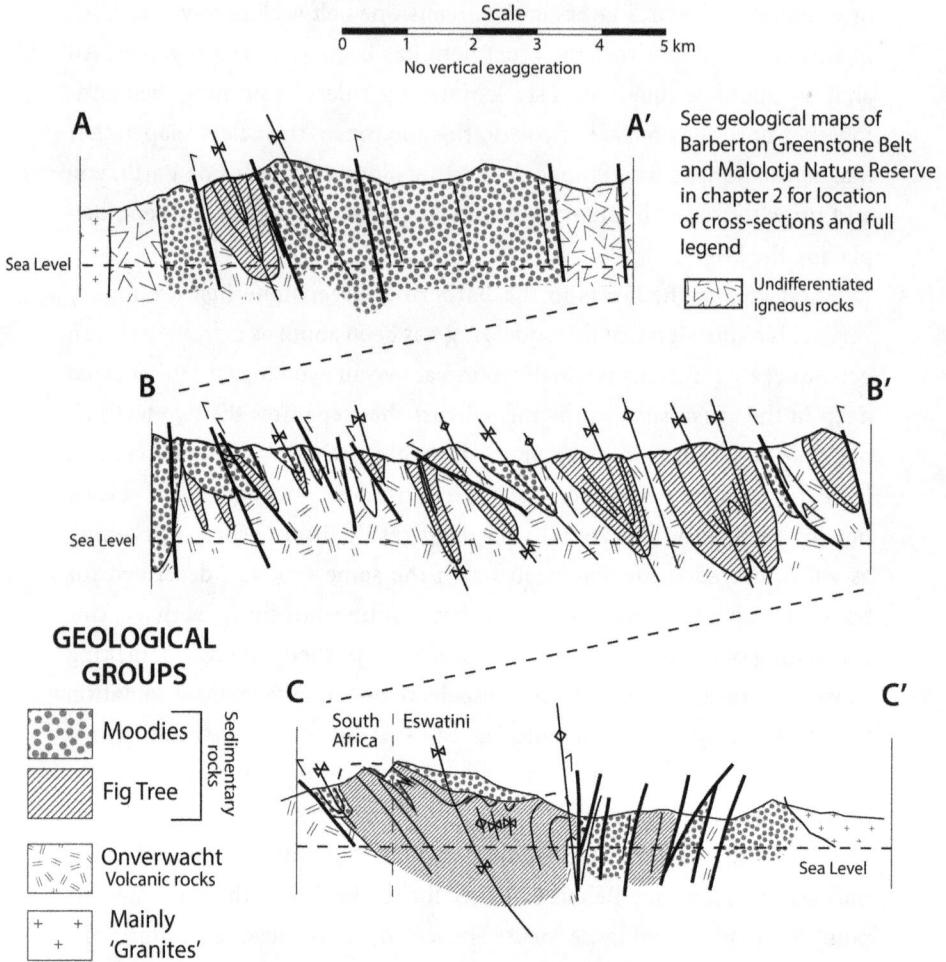

Scale

0   1   2   3   4   5 km

No vertical exaggeration

See geological maps of Barberton Greenstone Belt and Malolotja Nature Reserve in chapter 2 for location of cross-sections and full legend

Undifferentiated igneous rocks

GEOLOGICAL GROUPS

Moodies — Sedimentary rocks

Fig Tree — Sedimentary rocks

Onverwacht — Volcanic rocks

Mainly 'Granites'

FIGURE 5.3 Geological cross sections through the bedrock of the Barberton Greenstone Belt provide a sideways view— partly based on the 1: 25,000 scale geological map of the central Barberton Greenstone Belt by Don Lowe and colleagues—see the geological maps in chapter 2 for their location and full details. These reveal the attitude of the layers of sedimentary rocks, twisted into folds and cut by faults. The near-vertical orientation of most layers indicates intense horizontal compression, which thickened the crust and helped raise mountains on early Earth. (Illustration credit: Simon Lamb)

A FOLD IN THE LANDSCAPE

The Swaziland Geological Survey provided me with black and white aerial photographs of the Malolotja Nature Reserve. These had been blown up to a giant size so that I could map the rocks straight on to them. I had a bird's-eye view of the rocks where they outcropped in the landscape, and in areas where the grass had been burnt, I could pick out individual layers and trace them over the rugged hills for kilometers. One feature in the photographs that really caught my eye was the gigantic twist in the rock layers near the summit of the Ngwenya Ridge. Individual bands of strata formed a tight V, like the shape your leg makes if you bring your heel behind you to touch the back of your thigh. I have already mentioned this fold in the previous chapter, when talking about the iron ore mine at Ngwenya. It was those

FIGURE 5.4 The great fold at Ngwenya in the Malolotja Nature Reserve, Eswatini. Layers of Fig Tree and Moodies Group sandstones and conglomerates are twisted into a syncline. This fold was created by Earth movements that took place while rivers continued depositing sandstones and conglomerates across the landscape. In the distance is South Africa's flat Highveld, underlain here by granite bedrock. (Photo credit: Simon Lamb)

layers of sandstone and chaotic conglomerates in the Fig Tree Group, laid down in a submarine debris avalanche, that really made the fold visible in the landscape. They stood proud, like stone walls extending for kilometers, as though built by giants who couldn't care less about high hills or deep gullies. But there was something very odd about this fold—something it took me months of walking the ground to come to terms with.

Imagine folding a stack of paper—perhaps the pages in one of those thick telephone books—and then viewing it from the side. All the individual pages will have the same curved shape. If these were layers of sedimentary rock, a geologist would quickly conclude that the folding took place after the layers were laid down. But the Ngwenya Fold was not like this. You can simulate it more easily with a drawing. Start with a large, rounded V. Then draw another curve inside the V, slightly more open. Now repeat this process so that each new curve is even less pronounced than the last. Soon, you will have something that looks like the great fold at Ngwenya.

Yet the Ngwenya Fold was even stranger than this. The closest I can get is to suggest that, after drawing a few curves, you begin extending your lines at either end so they spill over as "wings," cutting across the ends of the previous V-shapes. In other words, the older layers in the Ngwenya Fold were abruptly truncated by younger ones. To a geologist studying sedimentary rocks, this is a dead ringer for what is called an angular unconformity. These form when the rock layers are first tilted, and then the action of erosion smooths and flattens the landscape. Subsequently, new layers are deposited on top, but they are no longer parallel to the underlying ones. At the unconformity itself, a span of time is unaccounted for; there is a gap in the geological record. Most angular unconformities are created in landscapes where rivers are eroding the bedrock. The younger layers at Ngwenya look just like those deposited by rivers: mainly sandstones and conglomerates, with thin mudstone layers. In fact, it is the thin mudstone layers that confirm this. They contain perfectly preserved mud cracks filled with sand, indicating episodes when the riverbed completely dried out.

My three years of undergraduate education had not prepared me for something like this. Again, it was Maarten de Wit who put me on the right track. He took one look at my geological map and declared that the only explanation was "synsedimentary deformation." "What on Earth is

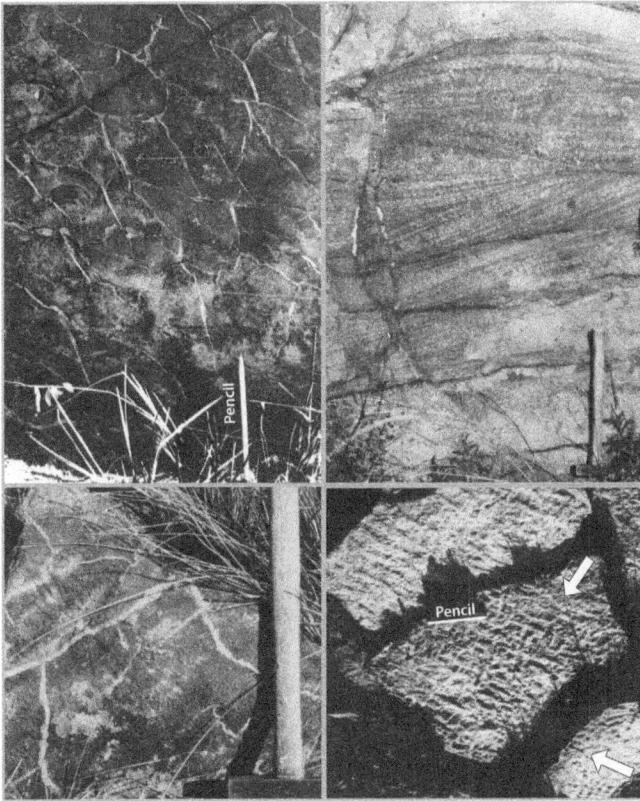

FIGURE 5.5 The rocks speak. Detailed features of sedimentary layers in the Moodies Group exposed in my study area in the Malolotja Nature Reserve, Eswatini, reveal clues about the rivers that deposited them. (*Bottom right*) Ripples on the surface of a layer of sandstone show river flow direction (see white arrow pointing to bottom left corner) and possible strong wind gusts creating new ripples at right angles (white arrow pointing to top left corner). (*Bottom and top left*) Polygonal mud cracks formed during low river periods, when mud dried out to be later filled with sand. (*Top right*) Underwater sandstone dunes with well-preserved internal cross-bedded layers of quartzite reveal water currents flowing right to left. (Photo credit: Simon Lamb)

synsedimentary deformation?" I asked. "You've got some reading to do, I think," he replied smugly. Back at Cambridge after my first major field season, I started asking around. It turned out that there was a research group in my department studying the geological record of the past 20 million years of rivers flowing off the Pyrenees in northern Spain. Some of these rivers had been active while the Pyrenees were still being pushed up, leaving behind an extraordinary record of geology in action at the surface. Older layers of sandstone and conglomerate had been tilted, as, geologically speaking, rivers flowed right over them—eroding and cutting into the layers in some places and depositing new ones elsewhere. All this had resulted in geological features that looked a bit like those I had seen in Swaziland.

I eventually came up with an explanation that seemed to account for the fold at Ngwenya. This is where Maarten's reference to synsedimentary deformation comes in. The prefix *syn-* just means "at the same time," as in *synchronous.* And deformation, to a geologist, refers to the distortion of rock layers, such as by folding or faulting. When applied to the Ngwenya Fold, synsedimentary deformation is a compact way of saying that folding took place when some of the rock layers at Ngwenya were being laid down *and* also while others were being eroded to create angular unconformities. But, as you will soon see, it was only later in my research career that I came to make full sense of it all.

MOODY MOODIES

It was not only in the Ngwenya Syncline that I found sedimentary rocks laid down by rivers. Other parts of my field area in the Malolotja Nature Reserve are also underlain by them. And similar sequences of sedimentary rocks are well known in the South African part of the Barberton Greenstone Belt. In fact, it was on the South African side that they were first described and named— collectively called the Moodies Group after one of the places where they are a prominent part of the bedrock, not far from the town of Barberton. The name is a corruption of Moodie's—that is, land belonging to a Mr. Moodie. As far as I can tell, Mr. Moodie was an early settler in the region. For simplicity, I will continue referring to all these rocks this way, although I originally gave them more local names based on where I found them in my field area.

Rocks in the Moodies Group are easy to spot, even when looking at the Barberton Greenstone Belt from space. This is because they are the bedrock to the main peaks and ridges in the Makhonjwa Mountains. These rise over 1,000 meters (3,000 feet) above the valley floors, requiring an exhausting climb up the steep and trackless grassy hillsides to reach the rocks. I think I must have climbed every one of them in the Malolotja Nature Reserve. And together with Maarten de Wit and my fellow student Isabelle Paris, I ascended many others. But once we reached the tops, the reward was wide open views of the Makhonjwa Mountains—a good place to sit and argue about the meaning of these rocks! In my field area, the rocky crags were also the favorite haunts of families of baboons, who made clear their displeasure at being disturbed with aggressive barking. But if I was patient, they would eventually move on, leaving the rocks entirely to me and my geological scrutiny. I would come armed with a notebook, long tape measure, and a geological compass so that I could measure the thicknesses of the layers and note down their particular features.

FIGURE 5.6 The author standing at the top of a spectacular waterfall in the Malolotja Nature Reserve, Eswatini, where the Malolotja River plunges into its deep valley. Immediately behind me are resistant ridges of Moodies quartzites. The skyline ridge on the left marks the international border with South Africa and is underlain by resistant, chaotic conglomerates of the Fig Tree Group. Just to the right of my head is Emlembe, the highest peak in the Makhonjwa Mountains, rising to about 1,800 meters (5,850 feet) above sea level. (Photo courtesy of Simon Lamb)

Back in the 1970s, the South African geologist Ken Eriksson was the first to study in detail the sedimentary layers in the Moodies Group on the South African side of the Barberton Greenstone Belt. He pioneered the use of modern geological techniques to work out how these were originally laid down, finding evidence not only for rivers, but also ancient shorelines and sand dunes shifted by the tides. I was particularly influenced by his work because it showed that despite the great age of these rocks, they could be read in the same language, as it were, as sediments deposited today. There is, however, one thing that is very odd about these rocks, which becomes obvious when you compare them to sedimentary rocks laid down by rivers in the much younger geological record. There is generally plenty of iron in river sediments, and where it has reacted with oxygen in the atmosphere, it has rusted, staining the younger rocks red and brown. When I was studying such sequences of sandstones in the Bolivian High Andes, I was perpetually in a landscape turned brownish red by this iron—the sandstones that built up the bedrock, and also the sandy beds of modern rivers, were this color.

In marked contrast, the sandstones in the Moodies Group are almost always a pale grey or whitish yellow—not just on their weathered surfaces, but deep inside. This is a sure sign that oxygen was essentially nonexistent in Earth's atmosphere at this time, which prevented the iron from rusting and then coating the sand grains with a red stain. To avoid confusion, I should say that the red iron-rich sediments of the underlying Fig Tree Group are not signs of an atmosphere rich in oxygen. This is because they formed out at sea, most likely, I think, through the action of algae (known as photoferrotrophs—see chapter 4) that are able to fix iron compounds in an environment free of oxygen gas. So, if you had wanted to walk about on dry land in this world, you would have needed oxygen tanks on your back.

BACK TO NEW ZEALAND

I came to a much fuller appreciation of the world in which the Moodies Group was laid down during my postdoctoral research in New Zealand. Soon after arriving there, I was invited to join a student field trip to the eastern part of North Island, in a region called the Wairarapa. The trip was led by Paul Vella, who had spent his career studying the rocks here. He had just completed a draft of a geological map of the region for publication by the New Zealand Geological Survey. Paul was anxious to ensure there were no obvious mistakes, so he went about double- and triple-checking everything. He delighted in taking colleagues to places where he felt he needed another opinion. One morning, while the students were busy with their own independent projects, Paul rather mysteriously asked me to accompany him to look at "something" he thought might interest me. As it turned out, this "something"—to my mind, at least—vindicated all my rather tentative ideas about how the fold in the ancient layers at Ngwenya had formed about 3.22 billion years ago.

Paul took me in his car and we parked by the Huangarua River bridge. After scrambling down to the river itself, we worked our way several kilometers downstream examining the outcrops of sedimentary rock in the riverbanks. At first, I did not find this particularly interesting, although I was in fact looking at rocks that recorded the emergence of this part of New Zealand from the sea about a million years ago. The rock layers were all gently tilted in the direction we were going, inclined at about ten to fifteen degrees. We began by examining some beautiful scallop shells. These shells were all jumbled up, and many were broken, indicating that this was no more than a rubble of shell fragments deposited in a shallow sea, cemented together to make a shelly limestone. The shells showed that the limestone was very young, geologically speaking, only a little over a million years old. Paul told me that beneath the limestone was mudstone deposited in much deeper water. As we struggled along the slippery riverbank, trying not to get our feet wet, the rock layers began to change. They were now sandstones, with only occasional shells in them, with thin beds of mudstone between the sandstone beds. Again, the fossils showed where these had been deposited—in the brackish water of a tidal estuary. We then found sandy beds full of pebbles, followed by thicker layers of conglomerate. The layers in these conglomerates were gently tilted, just like all the other layers Paul had shown me. And they had clearly been laid down by rivers less than a million years ago—probably a river just like the one we were following.

We had clearly left the sea and reached land in our geological journey along the banks of the Huangarua River. This is when I started to get interested. As we continued, we were still finding river gravels in the layers of rock, but the tilt of the layers was beginning to change. First the tilt gradually lessened, until the layers were more or less horizontal; then they began tilting in the opposite direction, inclined *upstream* toward us. We had obviously walked across a relatively open fold or warp in the rock layers—a syncline, in the jargon of geologists. The Ngwenya Fold in the Barberton Greenstone Belt is also a syncline—this just means that the layers become younger (the "way up," to geologists) toward the core of the fold. I was intrigued to see a fold like this in rocks at the other end of geological time, although I knew that New Zealand was in the viselike grip of a subduction zone, slowly squeezed between converging tectonic plates.

We were approaching the "something" that Paul wanted to show me. As we rounded a bend in the river, cliffs rose up on the right bank. These cliffs laid bare a spectacular view of all the layers of rock that we had passed over so far on our journey along the riverbank. But this time, instead of the layers being gently inclined, they were standing on end like books on a bookshelf. I was stunned to see such geologically young rocks tipped up like this. The same parts of the sequence that Paul had pointed out to me earlier on in our walk—marine mudstone and limestone, layers of sand and mud from the brackish water estuary, and consolidated river gravels—were all there. Along the top of the cliff, there were other layers of consolidated river gravels, inclined less steeply, resting on and cutting across the underlying rock bookshelf and forming an angular unconformity. The gravels *above* the unconformity could be directly traced into a sequence slightly upstream that was continuous with the rocks *below* it. It was this confusing arrangement of rock layers that had puzzled Paul. I think he was very surprised when I started talking about the Ngwenya Syncline. But I had immediately realized that I had seen all this before.

WRIGGLING RIVERS

I should now give you a feel for how rivers behave in this part of New Zealand. The Huangarua River snakes its way to the sea with numerous tight bends or curves, like a windy road through hilly country. The water in the outer part of each river bend has to flow faster to keep up with the

Restored cross-section through 3.22 Ga Ngwenya Syncline

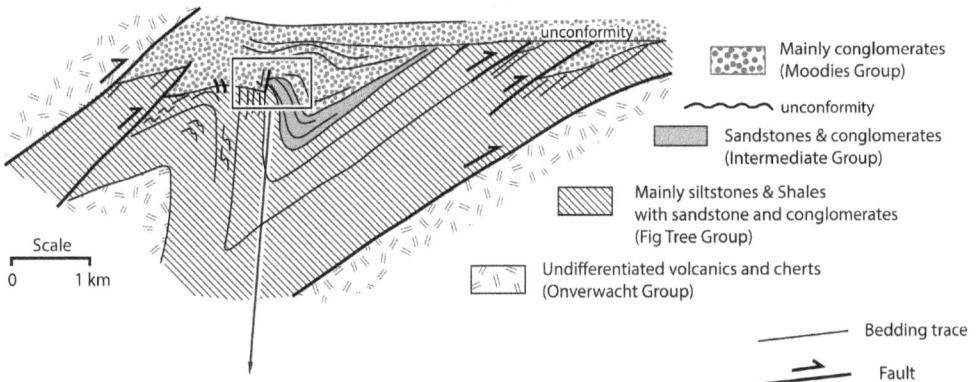

Mainly conglomerates
(Moodies Group)

~~ unconformity

Sandstones & conglomerates
(Intermediate Group)

Mainly siltstones & Shales
with sandstone and conglomerates
(Fig Tree Group)

Undifferentiated volcanics and cherts
(Onverwacht Group)

——— Bedding trace

———➤ Fault

Scale
0   1 km

Steep limb of 1 Ma Huangarua Syncline, New Zealand

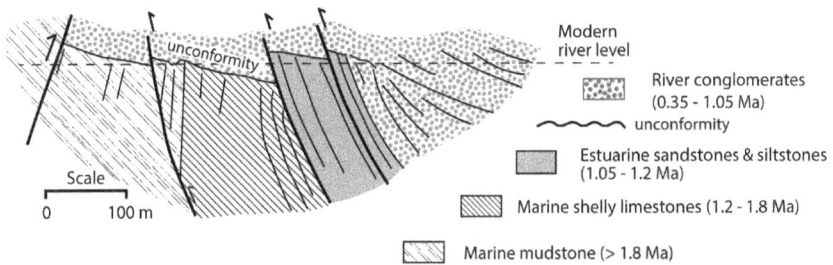

Modern
river level

River conglomerates
(0.35 - 1.05 Ma)

~~ unconformity

Estuarine sandstones & siltstones
(1.05 - 1.2 Ma)

Marine shelly limestones (1.2 - 1.8 Ma)

Marine mudstone (> 1.8 Ma)

Scale
0   100 m

FIGURE 5.7 The great fold at Ngwenya in the Malolotja Nature Reserve, Eswatini, as it might have appeared in Moodies times (about 3.22 billion [Ga] years ago), looking sideways on. The fold contains a detailed record that captures moments of time during its creation: while the sedimentary layers were being tilted and faulted, rivers flowed across them in a landscape that had recently emerged from the sea; in some places, these rivers deposited layers of sediment, but elsewhere they first cut into already tilted layers and then more layers were deposited on top to create an angular unconformity. These events are essentially the same as those that created the remarkably similar, but much younger, Huangarua syncline in New Zealand, which grew in the last million [Ma] years as the bedrock was uplifted out of the Pacific Ocean. (Illustration credit: Simon Lamb)

flow in the inner part. This increases the force of water, which tends to eat into the outer bank, especially during floods. Eventually, the river shifts its banks and the bend migrates. The overall effect of all this is that over tens of thousands of years the river wriggles over a wide swath of country, like some writhing snake, carving out the underlying bedrock to make a broad floodplain. As the climate switches between being wetter or dryer, hotter or colder, the rivers change their behavior too. Sometimes they build up the floodplain by dumping their cargo of debris. At other times they eat into the bedrock.

Over many cycles in the climate—especially if the land is being pushed up by forces deep within Earth—the rivers will tend to cut down through the bedrock and leave behind a series of river floodplains, called river terraces, like a giant staircase stepping down toward the present-day course of the river. This type of landscape is visible on either side of the Huangarua River. Scrutinizing the topographic map, it was clear to me that the terraces were not exactly flat but instead warped, mirroring the syncline I had seen in the riverbed. Progressively older terraces, higher up that staircase in the landscape, were warped to a greater extent. And looking even farther afield, it was clear that this was just one of many folds in the landscape of this part of New Zealand. Individual terraces had captured different stages in their growth, as though forming a series of snapshots or frames in a strip of celluloid film. Running all the frames together yields a smoother image in which the landscape is no longer stationary, but alive and moving.

The landscape and its underlying bedrock are not only warped into folds; they are also sliced up along fault lines that run wild through the region. One of these faults crosses the Huangarua River just downstream of the cliffs I was looking at with Paul. It was clear to me that movement on this fault was gradually nudging both the gravel plains and the bedrock into the fold we saw in the riverbanks. The best way to understand this is to think about what happens after a single earthquake. An earthquake is just the violent shaking that occurs when rocks suddenly break along a fault. Thus, during the earthquake there is some movement on the fault. The ground shifts upward—perhaps a few tens of centimeters (a foot or so). This drags up and tilts the bedrock, raising the riverbed near the fault but lowering it farther upstream, right in the heart of the bedrock syncline. The river soon smooths out its bed again, wearing away bedrock where it has been uplifted and adding gravel where it has dropped down. An analogy might be the way rail engineers construct a railway line over hilly terrain—by filling the valleys with ballast and bulldozing away the tops of hills.

If the same changes are repeated during subsequent earthquakes, perhaps every few centuries, a pattern emerges where some of the bedrock in the riverbed is worn away and other parts are built up with gravel. And during all this time the layers are being slowly tilted and warped. At some point, the river may change its way of smoothing out the riverbed, perhaps due to a shift in the climate or an adjustment in the uplift of the land. Now it builds up the bed everywhere by dumping gravel all the way along its course. In the railway analogy, engineers raise the ground level with ballast beneath the track so that it runs higher than any small hills. This is how I think the angular unconformity was created in the bedrock of the Huangarua River. Where there had previously been erosion, new gravel was deposited—but its thickness was slightly less than elsewhere.

I was fascinated by the fold that Paul had shown me. There were so many snapshots of its growth that I thought it would be possible to visualize its creation over geological time—certainly less than a million years—and to work out how fast it is growing today. And I soon did just this.

THE BIRTH OF A MOUNTAIN RANGE

My walk along the Huangarua River opened my eyes to events that occurred billions of years ago, allowing me to finally make sense of the Ngwenya Syncline in my old field area in Swaziland. This seemed to be an exact replica of the fold I saw in New Zealand—in shape, size, and the presence of unconformities. I realized that it must also have grown during emergence of the bedrock out of the sea, and all this could have happened in the same brief period of geological time as in New Zealand.

I think we have witnessed something much more profound in the warped sedimentary layers of the Moodies Group in the Ngwenya Syncline, and also the Moodies Group elsewhere in the Barberton Greenstone Belt: the birth of a mountain range. The uplifted seabed was being turned into a dry land traversed by rivers. And the headwaters of these rivers must have been in mountainous terrain, because, as I have already explained, the layers were being warped into folds and sliced up by faults, even as they were being deposited by rivers. These are all clear signs of mountain building, driven by the intense squeezing and thickening of the crust. We can work out the courses of the rivers through this terrain, because the shapes of the river dunes and ripples—fossilized in the sandstone layers as cross-bedding—tell us which way the rivers were flowing: the steep faces of the dunes point downstream. The rivers were flowing along the length of my study area, mainly toward the north. This direction runs more or less parallel to the gutter-like corrugations in the sedimentary layers, and typical of much younger mountain belts where major river valleys commonly follow the folds and faults in the landscape.

I imagine a landscape very much like the one I saw near the Huangarua River in North Island, New Zealand: a staircase of river terraces at the foot of mountains rising up a thousand meters or more above sea level—although,

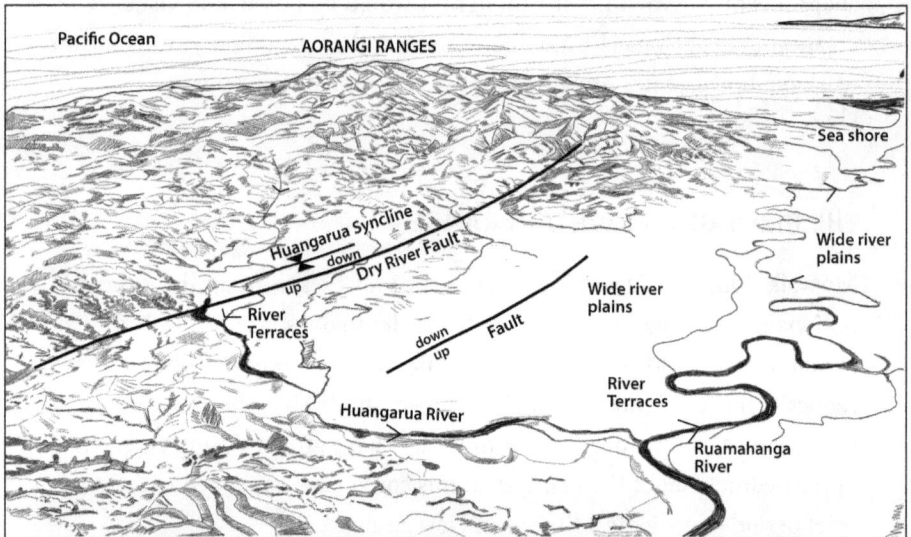

FIGURE 5.8 A sketch aerial view of the hilly landscape in North Island, New Zealand, looking toward the Aorangi Ranges. The Huangarua Syncline is actively growing today in the foothills of these ranges. During floods, sinuous rivers deposit sandstones and conglomerates, creating a flight of wide terraces as they also cut into the underlying bedrock. I think this landscape—minus the vegetation—resembles the one that existed during Moodies times, when the Ngwenya Syncline in my study area was forming. As in New Zealand, I think it was caught in the vise-like grip of colliding tectonic plates. (Drawing by Clara Maxwell Lamb)

of course, without the lush plant growth that is everywhere in New Zealand. The land surface of early Earth would have been mainly sand and gravel, interspersed with outcrops of naked rock. An important feature of the landscape on North Island is that it lies only about 100 kilometers, as the crow flies, from a deep oceanic trench where the Pacific Plate is sliding into the mantle along a giant fault in a subduction zone. Could the layers in the Ngwenya Syncline have been deposited by rivers in a landscape above a subduction zone too? I have already explained in chapter 4 why I think subduction was taking place when the underlying Fig Tree Group sediments accumulated. It therefore makes perfect sense for this subduction to still be active—the only difference is that by now the bedrock had emerged from the sea.

- § -

IN THE KITCHEN

We have reached the point in our time travels when there were mountainous tracts of dry land on the surface of early Earth. In my view, they were pushed up between two tectonic plates that were relentlessly moving toward each other in a subduction zone. Now is the time to return to the ideas put forward by George Airy in his revolutionary 1855 paper. As I explained earlier in this chapter, he had proposed not only an increase in the thickness of the crust to raise the level of Earth's surface, but also that the crust must have a lower density than the underlying mantle. The folding and faulting we see in the layers of rock in the Barberton Greenstone Belt certainly thickened the crust. But what sort of crust was it?

Today, the rocks here are clearly part of the bedrock of the African continent. However, as I showed in chapter 3, the early part of the geological record in the Barberton Greenstone Belt points to a very different place: the volcanic rocks in the Onverwacht Group erupted on the floor of an ocean, building up an oceanic crust far from land. And we know that oceanic crust is not the same as that beneath the continents, not only because the continental crust is thicker, but also because of the much wider variety of rocks, including abundant granites, compared to the monotonous basalts and komatiitic rocks in modern or ancient oceanic crust. In other words, at some point in the geological history of this region, continental crust was created. To me, the key question is whether this actually began when the sedimentary rocks in the Fig Tree or Moodies Groups were laid down and new tracts of dry land were emerging from the sea, eventually rising to form mountain ranges.

Our ability to answer this question depends on our wider understanding of how Earth's crust is created in the first place, because this will show us what we should be looking for. We can essentially boil it down to a matter of the chemical composition of the crust. To investigate this further, we need to revisit many of the ideas about the melting and crystallization of rocks I introduced in chapter 3, looking at how these ideas are connected to the composition of rocks. You may recall that geologists have traditionally gone about working out this composition by measuring the amounts of silica and other oxides they contain. I made these measurements myself for quite a few of the rock samples I collected in the Malolotja Nature Reserve. So, let me begin by describing what was involved.

My first task was to turn the rock sample into something that was, in effect, transparent. To do this, I first crushed and ground it into a fine powder. The powder was then mixed with a flux and heated to over 1,000°C in a furnace until it became a beautiful, glowing, and sticky liquid, rather like molten glass. The sticky liquid was then pressed into a glassy, colored disc—called a glass bead—which, when cooled, could be probed. I was exploiting the fact that x-rays can penetrate the semitransparent bead, exciting the individual elements inside so that they fluoresce and emit light with characteristic wavelengths. Using an instrument called an x-ray fluorescence (XRF) spectrometer, I could then measure the intensity of the fluorescence for a range of wavelengths, which is directly related to the chemical composition of the rock. When I was a graduate student, the whole procedure was very tedious and time consuming, taking several days for every sample; all this has now been automated, so that it is possible to analyze hundreds or thousands of samples in the time it took me to measure just ten. I used to sit for hours in front of the XRF spectrometer, writing down each number as it came up on the digital display while the instrument clunked through its various settings. I ended up with long lists of numbers detailing the proportions of the various oxides in my rock samples.

My time spent analyzing the rocks from my field area was a tiny part of a much wider research effort by many geologists to determine the oxide composition of rocks. To work out the composition of the continental crust itself, we need to pool all the measurements from individual rocks that are typical of the crust. If I consider just two of the oxides found in rocks, then putting all these rocks into a giant mixing bowl and stirring them together would yield continental crust with a composition of about 60 percent silica by weight and roughly 5 percent magnesium oxide (the rest is mainly aluminum oxide). For oceanic crust, the average composition would be about 50 percent silica and 10 percent magnesium oxide. By contrast, mantle rocks are only about 43 percent silica, but up to 40 percent magnesium oxide. It is obvious from this that the crust, compared to the mantle, is richer in silica—especially the continental crust—and poorer in magnesium oxide. The presence of those magnesium-rich komatiites and komatiitic basalts in early Earth's oceanic crust would have made it a sort of middle ground between modern oceanic crust and mantle.

So *why* does Earth have two types of crust, each with a different composition from that of the mantle? In chapter 3, I considered at length the origin of the oceanic crust in early Earth, which resulted in the production of basalts or komatiitic volcanic rocks. The key to this is either melting of the mantle in upwelling plumes of hot rock—rising all the way from the top of the underlying core—or melting of the mantle during seafloor spreading, helped by the addition of water from a sinking plate in a subduction zone. However, the formation of continental crust, with its average composition of about 60 percent silica, requires something more. This is really a question of having the right recipe and ingredients in a sort of geological kitchen deep inside the crust, because another way to describe the chemical composition of continental crust is as a mixture of the chemical compositions of oceanic crust and granite, in roughly equal quantities. So, in essence, we need to find a way of both making the additional granite and mixing it with basalt or komatiite.

THE SUBDUCTION FACTORY

An effective way to produce granite today is to let basaltic magma cool slowly and crystallize as it rises through the crust on its way to the surface. Crystallization, by incorporating or excluding particular components, alters the composition of the remaining magma. It is one of the consequences of magma getting trapped in the complex plumbing system below a volcano. Over time, the remaining magma is progressively refined until it eventually ends up with a granite-like composition. Mixing this with basaltic magma that has managed to rise much more rapidly through the volcanic plumbing system, so that there is no time to crystallize, creates a magma with the composition of average continental crust. This is also the way those distinctive volcanic rocks called andesites are created; rocks named after the Andes Mountains where they were first described. Andesites have a similar composition to average continental crust and erupt in the great arcuate chains of volcanoes that lie above subduction zones, such as the Ring of Fire around the Pacific Rim. Thus, the creation of andesites above subduction zones also builds up the continental crust. However, in the hotter interior of early Earth, the starting point in the granite refining process is a magma much closer in composition to the mantle itself—a komatiitic magma—and it is less likely to get far enough along the trajectory to granite. It turns out, however, that there are other ways to make the essential granitic ingredient in our recipe for continental crust.

Experiments show that in special circumstances it is possible to melt oceanic crust—made up mainly of basaltic or komatiitic rocks—to produce a magma with a very similar composition to the granites surrounding the Barberton Greenstone Belt. The catch is that the rocks need to be under the huge pressures that exist tens of kilometers below the surface, and they must have some water bonded inside them. Making granite this way requires *two* episodes of melting: first, melting of the mantle to make oceanic crust; and second, melting of the oceanic crust deep inside Earth with water as an added ingredient. The question then becomes: what went on in the history of early Earth that would allow the oceanic crust to melt like this? To my mind, the obvious answer, again, is subduction. If the oceanic crust sank at a subduction zone and penetrated the hotter mantle of early Earth, it would have eventually heated up enough under high pressures to melt—and the necessary water would also be on tap as sea water trapped in the rocks of the subducted sea-floor. The molten crust would have been buoyant enough to rise back up toward the surface and eventually crystallize as bodies of granite. Some of it may even have reached the surface as explosive volcanic eruptions.

Earth has other recipes, so to speak, when it comes to making continental crust. Volcanic rocks undergo a chemical reaction with the atmosphere—this is the process of weathering, which is the downfall of most inanimate things exposed to the elements, be they natural or human-made. But the minerals in the rock that are most prone to this process are those stable at depth—such as olivine, pyroxene, and feldspar—whereas quartz is highly resistant to weathering. Over time, the weathered minerals turn into mud and may be washed away back into the ocean. This leaves behind layers of

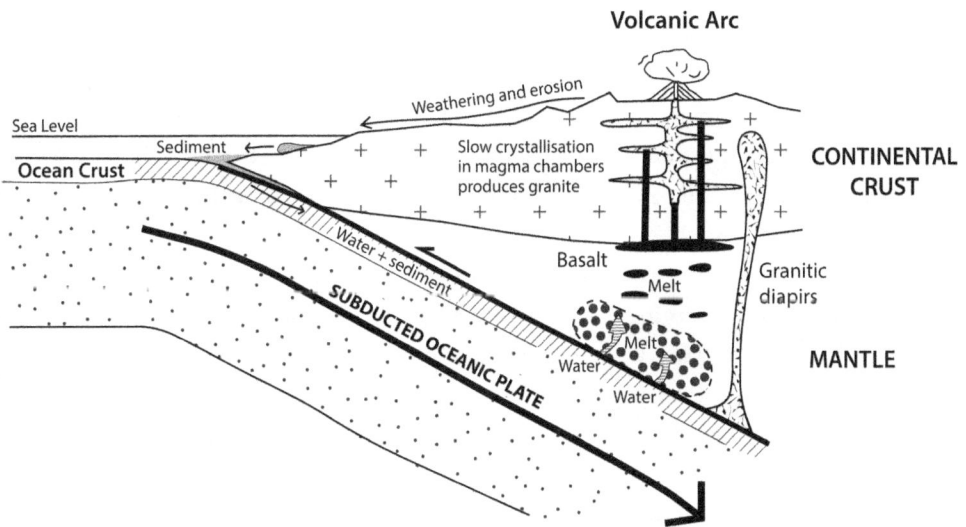

FIGURE 5.9 The subduction factory—how Earth makes continental crust above a sinking tectonic plate. The plate carries sediment and ocean water down with it, triggering melting in the oceanic crust and overlying mantle under the much hotter conditions at depth. The resulting magma rises as either pockets of basalt or diapirs of granite, slowly crystallizing in the overlying crust to form new granite or erupting at the surface in a volcano. Weathering and erosion at the surface complete the rock cycle. (Illustration credit: Simon Lamb)

sediment that are mainly quartz. I have seen this for myself in parts of South Island, New Zealand, where the undulating landscape is strewn with fragments of quartz left behind as the other, more weathered rock has been winnowed out by wind and rain. Thus, the creation of dry land can itself help to change the composition of the crust by providing another way to refine its silica content, especially if the climate is warm and wet.

We can now try to put together the story of how the continents were made. Most likely, there was very little continental crust to begin with in our planet's history. Rather, the crust was made of komatiite, komatiitic basalt and the usual basalts, and it lay beneath deep oceans where it was penetrated by water. Eventually, the water-rich crust started to sink back into the mantle, and here it experienced high enough temperatures in the hotter mantle of early Earth to melt, producing a granitic magma. There was so much water around that it also triggered melting of the surrounding mantle—like adding antifreeze—producing basaltic or komatiitic magma, which could then begin to be refined in the plumbing systems of the overlying volcanoes, crystallizing as more silica-rich rocks. Either way, production of continental crust had started. I think all this happened in a subduction zone, and if there were periods when the plates moved much faster, with a greater rate of subduction, then there would be more production of this crust.

In time, as more continental crust was produced and it was squeezed between the converging plates in the subduction zone, it emerged above sea level to form dry land. More dry land increased the exposure and weathering of rocks, leaving behind a wider area of quartz-rich detritus. If this weathered sediment were then carried to the sea by rivers and transported into an ocean trench as turbidity currents, it could itself be subducted and melted to make more crust. And because continental crust is less dense than the underlying mantle, it tends to remain floating on top of the mantle. Once created, the continents are essentially here forever, though their thickness and configuration may change over time.

It seems I have just described what sound engineers would call a feedback loop, capable of amplifying itself. Geologists sometimes refer to this way of manufacturing crust as the subduction factory. And like any factory, as early Earth's production facilities were scaled up, so was its output.

Once continental crust existed, it could be squeezed and thickened to further raise the level of Earth's surface, creating large tracts of dry land and mountain ranges.

THE INCREDIBLE LIGHTNESS OF BEING

We can observe in the Fig Tree and Moodies Groups all the signs of the subduction factory in full swing, producing the key components of continental crust such as granites and other silica-rich rocks. In the Fig Tree Group, there are thick piles of volcanic rock with a composition of about 65 percent silica, typical of the lavas that erupt in the volcanic arcs above subduction zones. These lie in the regions north and west of where Isabelle and I were working, but they have been extensively studied by an American team of geologists led by Don Lowe at Stanford University. There are also indications in the layers of sandstones and conglomerates in the Fig Tree and Moodies Groups that granite was becoming an important rock type at this time.

I discovered this myself when I started looking closely at the samples of sandstone I had collected in the field. To do this, I had to turn the rocks into wafer-thin slices—what geologists simply call thin sections—which are transparent to the naked eye. By shining light through the thin section with a polarizing microscope, I could look right inside the individual crystals and identify the particular minerals, extracting a surprising amount of information about how they formed and their subsequent geological history. During my project, I spent many hours with my eye glued to the eyepiece of a polarizing microscope, twisting the thin section round and round on its rotating stage in order to find the perfect lighting angle. I wanted to see the individual grains of sand that had been plucked by rivers from the bedrock of the uplifted landscape and then deposited as layers of sediment. This way, I could work out what this bedrock was.

I found that the older sedimentary rocks in the Fig Tree Group were mainly made up of fragments of chert eroded from the underlying Onverwacht Group, or cannibalized from the Fig Tree Group itself. This was easy to see, because under the polarizing microscope the individual sand grains were a mosaic of minute crystals of quartz welded together—called polycrystalline quartz—typical of chert. But near the top of the Fig Tree Group, and in the river-deposited sediments of the overlying Moodies Group, a profound change had taken place: many of the sand grains were now just large single crystals of quartz. I think granite or other silica-rich and coarse-grained igneous rocks (which I also loosely refer to as granites) are the only plausible source of these grains, because they are made up of large interlocking crystals of feldspar and quartz. At Earth's surface, the feldspar weathers away, but the quartz survives. Indeed, it is just this sort of weathering that I described earlier on when discussing ways to make continental crust. And there seemed to be no shortage of granites, because some of the sandstones contained almost nothing else but crystals of quartz, making the rock a quartzite.

Clearly, the headwaters of the rivers at this time were eroding a bedrock of granites, perhaps forming rugged high peaks. I think that much of this granite was manufactured in the subduction factory, eroded almost the moment it came off the production line. But some was clearly part of the bedrock of older continents, such as the roughly 3.6-billion-year-old gneiss and granite exposed in the riverbanks on the edges of the Makhonjwa Mountains—fragments of gneiss and granite have also been found in some of the conglomerates of the Moodies Group. These continents must have been far away when the volcanic rocks in the Onverwacht Group erupted on the floor of an ocean. So, by some means, they were subsequently brought closer together—and again, I think this was due to subduction of the intervening ocean floor. This ultimately resulted in the collision of the continents, welding them together to form an even larger landmass. Visualizing the new landscapes created at this time is the next stage in our quest. Again, the answers lie in the rocks of the Moodies Group.

COLLIDING CONTINENTS

You may remember my first trip among the sedimentary rocks of the Moodies Group, which I recounted at the end of chapter 2. At that time, I was confronted with the problem of working out the "way up" of the sandstone layers. After a long search, I solved this by finding unmistakable

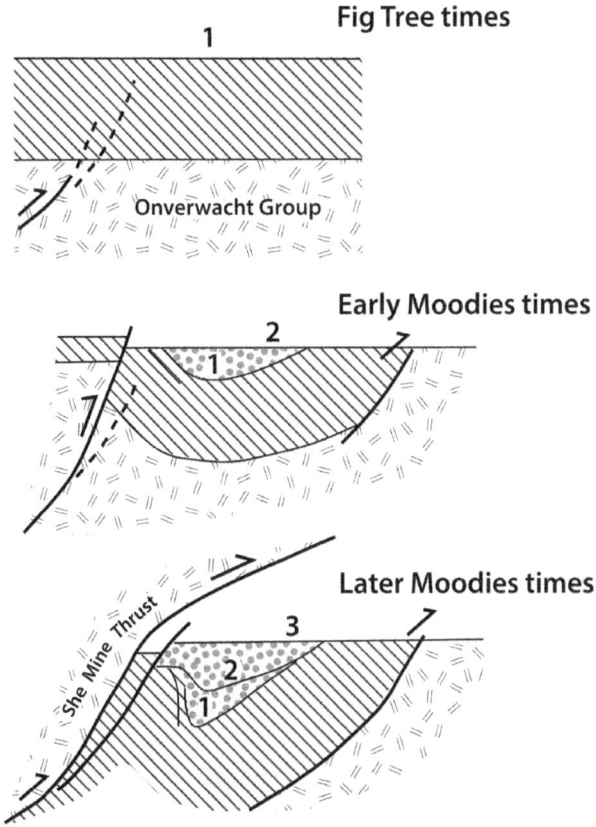

FIGURE 5.10 A collision in action. A hypothetical diagram illustrating side-view stages in the growth of the Ngwenya Syncline in my study area during Moodies times, showing how it was connected to movement along a major fault—the She Mine Thrust—which shoved older Onverwacht Group volcanic rocks over younger layers. During this time, rivers either eroded the underlying bedrock of Onverwacht and Fig Tree Group rocks, or laid down new layers of sandstone and conglomerate on top. The entire process may have resulted from the immense horizontal forces generated when tectonic plates collide. (Illustration credit: Simon Lamb)

cross-bedding in the sandy layers. This revealed a giant fault that had uplifted komatiite lavas of the Onverwacht Group, placing them directly on top of the sandstones—the komatiites must have been pushed along the fault for several tens of kilometers or more. Once I had recognized this fault, I found several others, and movement along these faults seemed to have stacked the rock layers like a shuffled pack of cards. The scale of all this is what you would expect when continents collide, because continents are too buoyant to be subducted, and so the two plates face each other head to head, as it were, squeezing everything in between like giant buffers at the end of a railway line, while the remnants of the oceanic plate break off as they continue to sink deep into the mantle.

We need also to take into account that the Moodies Group does not only contain evidence for rivers, but also a shoreline and shallow seas. Some of the quartzites exposed in my study area in the Malolotja Nature Reserve contain evidence for dunes and ripples moved by currents that periodically reversed themselves. This is typical of sandstones that are deposited during reversals of the tide and the back-and-forth motion of waves close to shore. Ken Eriksson, in the late 1970s, was the first to describe sandstones like these from the Moodies Group across the border in South Africa. He showed that the ancient tidal rocks look no different from those forming today, suggesting that tidal forces on early Earth were similar to today's—I will talk more about this in chapter 7. But I also think they show us that the great weight of the newly formed mountains was pressing downward on the edges of the collision zone, as is commonly observed in much younger mountain belts, creating a depression that was flooded by the sea. I like to think of something like the Persian Gulf today, where the weight of the Zagros Mountains has depressed the edge of the Arabian Peninsula, forming a seaway in the collision zone between the Arabian and Eurasian plates, in the region immediately southwest of modern-day Iran. The northern end of the Persian Gulf is formed by the delta of the Tigris and Euphrates rivers, which drain the interior of the Middle East. Here, one finds a transition from river-deposited sediments on land to tidally influenced sediments offshore in a depression between the colliding continents.

I should say here that there is no scientific consensus on the interpretation of the Moodies Group in terms of plate tectonics, and some geologists see no need to invoke plate tectonics at all. I will leave the discussion of their version of events until the end of this chapter, after we have looked more closely at the surrounding granites. To me, however, the evidence clearly shows that two continents were brought ever closer together in a subduction zone, with subduction of the intervening ocean floor, until

MOODIES times (about 3.22 Ga)

Continental Collision

Crust of komatiites, komatiitic basalts & basalts

Fig Tree sediments

Moodies sediments

'Granitic' crust

'Granite' diapirs

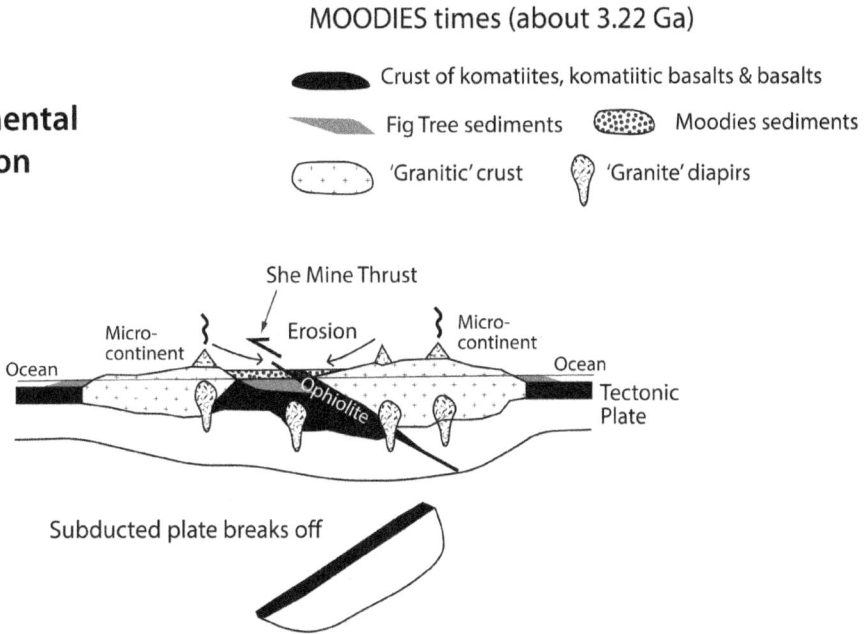

FIGURE 5.11 Hypothetical diagram illustrating the final stages in the history of the Barberton Greenstone Belt during late Moodies times, showing the collision of two small continents. The subducted ocean plate has broken off and foundered. In this reconstruction, the Onverwacht Group forms an ophiolite, typical of much younger mountain belts, and continued volcanic activity is driven by the intrusion of granite bodies and diapirs, which are subsequently eroded when they reach the surface. The major faults and folds in my study area, such as the She Mine Thrust, are due to the final squeezing of the crust in the collision zone. (Illustration credit: Simon Lamb)

they eventually collided. The continents were finally welded together by movement along the giant faults I found in the Malolotja Nature Reserve. This created an ophiolite belt from the relics of seafloor in the Onverwacht Group, much like those narrow, sinuous belts of uplifted oceanic crust that snake through many of the world's great mountain ranges. In the closing act of this collision, the rocks were squeezed still further, folding even the major faults themselves. It was this that sealed the fate of the Barberton Greenstone Belt. One continent is now part of the bedrock of Eswatini to the east, while the remains of the other lie in South Africa, on the western and southern flanks of the Makhonjwa Mountains. Together, they form part of the jig-saw puzzle of ancient continental and oceanic fragments that is the Kaapvaal Craton, assembled through the motion of the tectonic plates.

We can establish a timeline for the final collision of the continents, and it turns out to be very tight. This was first worked out by my New Zealand colleague, Cornel de Ronde. He found granite-like igneous rock that had intruded into the Fig Tree Group layers *before* the Moodies Group sandstones and conglomerates were laid down on top. He managed to date these intrusions with remarkable precision, using the time capsules provided by those tiny zircon crystals, which appear to faithfully record geological events. The age Cornel extracted was 3.227 billion years, plus or minus 3 million years. This is the same age, within the margin of error, that Cornel got for granites at the edge of the Makhonjwa Mountains. But these granites were latecomers, invading their way into the greenstone belt *after* the Moodies Group had already been deposited.

Cornel's new dates showed that the geological events recorded by the Moodies Group occurred over a shorter time span than can be clearly resolved for this great age. This short timeline has been confirmed by Chris Heubeck of the University of Jena, using high-precision dates for volcanic rocks within the Moodies Group itself. He showed that its deposition and subsequent folding and faulting occurred over a time span that may have been as short as 1 million years, but was certainly less than 14 million. We have to turn to much younger rocks, such as those along the banks of the Huangarua River in New Zealand that I described earlier—folded into a syncline just like the one at Ngwenya in Swaziland—to convince ourselves that this geologically short time span is sufficient for such great upheavals of Earth's surface to run their course.

Today, large tracts of the bedrock around the Barberton Greenstone Belt are composed of granites that were once molten about 3.1 billion years ago, and roughly 0.1 billion years (that is, 100 million years) after the deposition of the Moodies Group. I think that much of this granite was the product of the final paroxysm of melting of the thick crust formed during continental collision. Once it had cooled and solidified, a new continent emerged, making up the core of the Kaapvaal Craton. The forces of erosion soon began to wear away the highlands, and the detritus was washed onto a wide plain where it accumulated as thick piles of sedimentary rock, exposed today in the Witwatersrand on South Africa's Highveld, around the city of Johannesburg. And in this detritus lay a fantastic mineral wealth of gold, yielded up by the ancient greenstone belts—gold that nineteenth-century prospectors would later find as they explored this region. That, however, is a story for the next chapter.

ROCK PIERCER

We still need an explanation for the distinctive shape of the Barberton Greenstone Belt. You may recall my analogy involving a cookie cutter to describe how it looks like the wispy leftover pieces at the edges of the more circular granite "biscuits." If you look closely, you will see that the greenstones have been cooked up near these edges and individual minerals have recrystallized into new forms that are stable at the higher temperatures of the once molten granite. Here, in both the granites and the greenstones, individual crystals that make up the rocks have a plate- or wafer-like form, creating numerous planes of weakness along which the rocks tend to cleave or fracture. The crystals are stretched out too, so that they resemble flattened and elongated pumpkin seeds. This distortion of the crystals shows that the granite, even after it had solidified, was still capable of flowing over geological time. But there is more.

Grains of sand in the Fig Tree or Moodies sandstones, or even boulders in the conglomerates, have also been squeezed and remolded. These features are particularly prominent at the southern end of my study area, near the granites. I was very confused when I first encountered this phenomenon, because what I saw depended very much on which way I looked at the rock. Imagine examining a rugby ball from different angles: viewed from the side, it is a flattened ball; viewed end-on, it is completely circular. It was a similar problem with the sandstone grains. Viewed from the side of a rocky outcrop, they were stretched out as rods, making the rock from this angle look a bit like streaky bacon. But from the top, the streaky pattern disappeared and it was possible to recognize more typically shaped sand grains. I started to collect samples and make measurements of these different shapes, carefully noting the viewpoint. I assumed that when the sandstones were originally deposited, the grains appeared roughly the same from all angles.

This way, I was able to work out not only how the rocks had been squeezed and stretched, but also the fact that the distortion of the grains became less and less marked for rocks found farther into the interior of the greenstone belt: more than a few kilometers away from the granites, the squeezing and stretching were virtually undetectable. A simple explanation for this is that the rocks near the edges of the greenstone belt had been softened by their proximity to the once-molten, still-hot granite, allowing them to flow and be molded into new shapes as the granite—after solidifying—continued to push upward and intrude further into the greenstone belt. As an analogy, think of how plasticine molds and stretches itself around your finger when you push your finger deep into a lump of the stuff. Here, your finger is acting like the rising solid granite.

Geologists use the term *diapir* for a body of rock that behaves in this way, forcefully pushing through the surrounding rocks. The word comes from

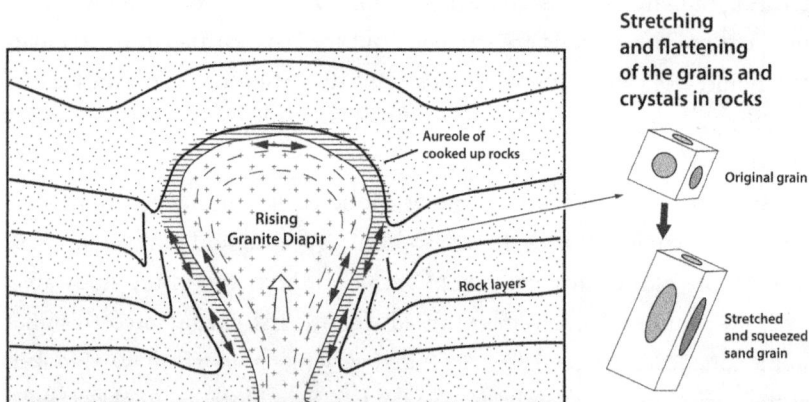

FIGURE 5.12 Rock piercer. Rising masses of buoyant granite form diapirs that push through overlying rock layers, twisting them into folds and squeezing and stretching individual crystals or sediment grains. This sketch is based on a geologically young diapir in China that rose about 130 million years ago, but it has all the key features of granite diapirs found at the edges of the Barberton Greenstone Belt—formed over 3 billion years ago. (Illustration credit: Simon Lamb)

the Ancient Greek, meaning "to pierce through." Rather like the upwelling blobs in a lava lamp, which I used as an analogy for mantle plumes in chapter 3, diapirs are driven by the density contrast between the less dense granites and the overlying, denser ultramafic rocks of the Onverwacht Group. In this case, the diapirs formed within the underlying low-density granites. These diapirs must have been one of the last geological events in the greenstone belt, remolding the volcanic and sedimentary rocks and creating the distinctive cuspate margins. Blair Schoene at Princeton University and Cristiano Lana at the University of Stellenbosch have managed to confirm my late timetable by directly dating when the individual crystals in the rocks changed shape, using the radioactive clock.

The differences in density between the rocks in the Barberton Greenstone Belt and the surrounding granites have another, more practical consequence. They make it possible to estimate how deep the greenstone belt extends below the surface. This is because both the density and *volume* of the rocks in the greenstone belt will influence the local pull of gravity, and it is possible to measure this with a very sensitive instrument called a gravimeter. As part of my project, I analyzed the results of this work using a computer program given to me by my thesis supervisor Alan Smith. He had written it back in the early 1970s when some computer programs still consisted of a stack of punch cards. Despite carelessly dropping and mixing up the stack, I did eventually manage to get it to work. The results clearly showed that the Barberton Greenstone Belt lies only "skin-deep" in the crust, underlain by granite at depths of no more than about five kilometers. The greenstone belt is effectively enclosed from the sides and below by the diapirs of granite, like an insect trapped in amber.

Similar behavior of granites is thought to occur in the thick continental crust of much younger mountainous regions, such as the Andes or Himalayas, where there is evidence for large bodies of once-molten granite. The crust has been squeezed by the forces that drive the converging tectonic plates. The resulting thickened crust slows the upward progress of the magma from the underlying subducted plate, giving it time to crystallize and solidify as granite. The crust may also be sufficiently hot at depth for the surrounding rocks to melt locally, crystallizing into yet more granite. And all these rocks are hot enough to flow, capable of pushing upwards as buoyant solid diapirs and eventually engulfing rocks nearer the surface. In this way, and ultimately as a consequence of the thickened crust created by plate tectonics, a fragment of Earth's surface could become embedded in granite and preserved for billions of years.

Some geologists, however, see the granites in a very different way. In this view, the rise of the granites is not a late-stage product of plate tectonics, but rather one of the fundamental driving forces of early Earth. Martin Van Kranendonk at the University of New South Wales has returned to a concept that, to me, is not too dissimilar from those gregarious batholiths described in chapter 2, originally proposed by Alexander Macgregor in the 1950s. He sees the Onverwacht Group in the Barberton Greenstone Belt as the product of mantle plumes that triggered prolific outpourings of lava. Eventually, the deeper parts of the volcanic pile began to melt, creating buoyant granites. From then on, the geological history was controlled by the forceful movement of these buoyant granites, which rose up towards the surface as diapirs and displaced the overlying denser rocks. All this activity caused profound changes in the level of Earth's surface, first by creating the conditions for the deposition of the Fig Tree and Moodies Groups, and then by squeezing the rock layers to produce the faults and folds we now see in the greenstone belt.

Clearly, whether or not you think plate tectonics had a role in all this depends on what you consider to be cause and effect. In my view, as I have argued throughout this narrative, plate tectonics is a fundamental cause of what we see in the Barberton Greenstone Belt. I will develop this argument further in chapter 7, but this does not mean that plate tectonics was the only game in town.

- § -

LIVING APART

I would feel dishonest in this account of my time in southern Africa if I did not face up to the realities of Apartheid. Apartheid is an Afrikaans word that sums up the political policy of keeping the melting pot of ethnic groups in South Africa apart. This could extend to the smallest details of everyday life, such as the use of buses, park benches and public toilets. The people who ran the show—and who wanted to be apart—were the Whites (Blankes in Afrikaans), mainly of Dutch origin and speakers of a Dutch-derived language called Afrikaans. Everybody else was crudely labelled as non-Whites or non-Europeans (nie-Blankes), which, of course, referred primarily to Black South Africans. Like all attempts to institutionalize racial policy, it was a ghastly system, rightly condemned by the rest of the world.

When I first went to southern Africa, I was very ignorant of all this. My focus was on being a geologist, studying the ancient rocks in Swaziland. But to get to these rocks, I had to travel though South Africa from Johannesburg. Eventually, I visited many parts of South Africa and neighboring countries, including Namibia, Botswana, and Zimbabwe—and so I came face to face with the realities of Apartheid. Let me say that nobody I met was happy about the Apartheid regime, and they all wanted to be free of it. Maarten, in my view, was one of the bravest, and he spoke out about it all the time to anybody who would listen. We used to have intense and emotional conversations in which we proposed what we thought were solutions.

But we were shielded from the true horror. All around us were millions of people who lived their lives as victims of Apartheid. One of the many tragedies of Apartheid was that it not only wasted lives, but also it wasted so much time in the development of southern Africa. Anyone who was not white was generally denied access to a good education. So when South Africa finally broke free in the early 1990s, the country faced the enormous task of catching up and cultivating a significantly educated population to run the nation effectively. Maarten's response was to become heavily involved in helping universities improve their capacity to admit more Black South Africans, who were starting from a much lower educational baseline than white students.

The official policy over Apartheid was that South Africa was in effect two separate countries. There was "White" South Africa, comprising most of the land that was valuable, either for farming or mining; and there was "Black" South Africa, a hodgepodge of disjointed and convoluted homelands. But there was very little employment in the homelands, and White South

Africa relied on the large cheap labor force provided by the inhabitants of the homelands. This meant that the Black community had to have special passes in order to "visit" White South Africa. These "visitors" lived in the townships around big cities like Johannesburg, or in dormitories provided by the mining companies. The situation was far more complex than this, with additional groupings under Apartheid, such as the Cape Coloureds, Bushmen, and Indian populations in the southern and western Cape provinces. My experience, however, was primarily in the northern part of South Africa, around Johannesburg.

What all this meant in practice was that if you were a white person in White South Africa, you were only likely to meet Black South Africans who were working for you, either as domestic staff or as workers in small businesses or factories. I felt that those I did meet were afraid to say what they really thought, and so it was difficult to engage in a conversation other than on very general topics. But I found that when I said I was working in Swaziland, I was rewarded with a wide smile. At that time, Swaziland seemed to be the land of the free, and I always breathed a huge sigh of relief when I got through the road checkpoints at the border and was truly out of South Africa. There was no doubt that South Africa was a very tense place, especially as the guerrilla war with the African National Congress (ANC) intensified. To me, Swaziland was the opposite: a bustling, confident and relaxed country.

One of the astute policies of King Sobhuza II of Swaziland was to maintain neutrality by not overtly aligning Swaziland with Black Africa to the north, or harboring ANC fighters. The South African Defense Force (SADF) kept a careful eye on the border, especially where it ran through the rugged country of the Makhonjwa Mountains on the edge of the Malolotja Nature Reserve. I was warned that a trickle of ANC fighters crossed the border from Swaziland into South Africa, but the SADF preferred to keep tabs on them and see who they interacted with, rather than challenge them at the border. A lurking fear I had was that I would be targeted by either side, especially as I was driving around in a Land Rover so close to the border. But I was left alone, and I never saw anybody. I think that I may have been protected by the presence of the reserve, and perhaps it was assumed I was a park ranger and not worth bothering.

There were international aid organizations operating in Swaziland, including the United Nations (UN) and the American Peace Corps. I got to know a number of Peace Corps volunteers; we would meet up on weekends when I was taking a break from fieldwork and they were coming back into town. The Peace Corps, established by President Kennedy in 1961, had a policy of integrating its volunteers into the communities that they were helping. In practice, this meant the volunteers had the same standard of living as the rural Swazis, and they were not provided with vehicles unless there was a special reason. One of these volunteers used my motorbike, and I think this made his life much easier. My Peace Corps friends identified closely with the communities they were working with, and they had all learned Siswati. Some had local girlfriends. To me, the Peace Corps was one of the very best forms of foreign help. Certainly, the UN specialists and workers from other aid organizations I met, who were effectively diplomats and had high salaries and living allowances, operated in quite a different world. But I suspect they all found it hard to see where I fit in. They were presumably making a difference to the lives of the Swazi people. So, what difference was I making? This is a question I am still trying to answer—especially now that I have worked as a geologist in many parts of the world where, in terms of material resources or economic wealth, the people I met were much worse off than myself.

Interlude

6 | King Solomon's Mines

The young planet yields up its mineral wealth

The exploration of our planet's bedrock has gone hand in hand with efforts to exploit its mineral resources. It was the old prospectors, looking for gold back in the late 1800s, who were the first to alert survey geologists to the existence of the remarkable rocks in the Makhonjwa Mountains of South Africa and Eswatini. These prospectors kept their ears close to the ground, listening out for rumors of any chance find. From wide experience in searching, they seemed to have developed almost a sixth sense that allowed them to quickly home in on rich lodes of gold hidden in the vast expanses of the newly established colonies. Their main technique was to look for signs of gold in the sandy beds of rivers—revealed by the telltale glittery residue left behind after swirling the sand in a pan—and to follow the evidence up stream until they hit the bedrock source. Since this early exploration, vast amounts of mineral wealth—not just gold, but many other minerals too—have been discovered and mined. During my third field season, I saw for myself some of the big mines in the regions west of the Makhonjwa Mountains, in the Highveld around Johannesburg, and even farther west in Namaqualand and Namibia. This way, I gained firsthand knowledge of how the rocks in my field area fit into the much bigger geological story of southern Africa, and how this story is intimately bound up with the planet's mineral resources. It was certainly fascinating to see the way these resources were being extracted and talk to the mine geologists about their operations.

The opportunity to visit these mines occurred rather unexpectedly. I had been visiting Isabelle, my fellow student in the Makhonjwa Mountains, in her field area; we were collaborating on a unified description of the rocks. She was staying with the family of the manager of a local mine, who lived in a comfortable house close to her field area. The garden had large jacaranda trees with cascades of purple flowers, making an almost idyllic scene against the backdrop of the high ridges of the Makhonjwa Mountains. I camped in the garden, but Isabelle was treated as one of the family and had a room in the house. She told me that she was planning to go back soon to Wits— the University of the Witwatersrand in Johannesburg—to join a student geological trip to Namaqualand and Namibia as part of their economic geology course. It was organized by one of her colleagues.

I remember feeling very envious, and I asked if I could join the trip. She telephoned Wits and left a message about my request. It seemed that it all depended on how many students had registered. Eventually, I heard back that there was room for me. But by then I had decided I could not afford to take so much time out of my field season. When Isabelle's host—the mine manager—heard that I had changed my mind, he revealed his intense exasperation. I think he had had enough of listening to me agonizing about what I should do. And in the end, I did go. In this chapter, I want to take you on the same journey of discovery.

A HIDDEN TREASURE TROVE

I have to admit that, prior to my trip, I only had a very sketchy knowledge of the mining industry. I certainly did not appreciate the scientific side of prospecting and mining a mineral deposit; I thought of mineral deposits as little more than lucky and easily dug up discoveries by adventurers or intrepid explorers. In the back of my mind were the adventure stories I had read as a child, particularly Rider Haggard's *King Solomon's Mines*, which was first published in 1885 around the time prospectors were flocking to southern Africa. The story is also set in southern Africa, although it takes place in a composite landscape that may even include the Makhonjwa Mountains. The plot involves a group of explorers in search of the fabulous mythical wealth of King Solomon. This turns out to be a treasure house of gold and diamonds, hidden deep inside a mountain in an unexplored region, and sealed by great slabs of stone. It is also guarded by an army of hostile locals. To quote the book: "'Lad, did you ever hear of the Suliman Mountains up to the north-west of the Mushakulumbwe country?' I told him I never had. 'Ah. Well,' he said, 'that is where Solomon really had his mines, his diamond mines, I mean. An old Isanusi or witch doctoress up in the Manica country told me all about it.'"

I still see a nugget of truth in Rider Haggard's tale, because, as I have already mentioned, there really is fabulous mineral wealth hidden in remote parts of Africa—this is not mythical at all. However, the story also highlights for me a fundamental misconception about this wealth: it is not just diamonds and gold—although there is no shortage of these in Africa—but more importantly, deposits of less glamorous minerals, such as the metals with which we have built up our modern world. I think the origins of these deposits are even more exciting than the premise of *King Solomon's Mines*. We don't need to thank a long-dead and mysterious king for conveniently stockpiling them in one place. Instead, we can thank the geological events that took place mainly in the first half of our planet's history—going all the way back to when those rocks in the Makhonjwa Mountains that form the Barberton Greenstone Belt were laid down at the Earth's surface. The ultimate source is Earth's mantle, deep beneath the crust, brought up by volcanic activity. Most people are only familiar with human-made end products, such as jewelry, copper wires, batteries, smartphone touch screens, computer microchips, metal car bodies, steel girders, or nuclear reactor fuel. So, let me introduce you to the biography of the mineral deposits themselves, from which these things are derived.

There is one word that sums up a mineral deposit: concentration. It is this concentration that creates a natural subterranean treasure trove.

Sometimes, the concentration is a one-step process. The world's largest deposits of chrome, nickel, and platinum-group elements—key components of high-performance metal alloys, electronics, and batteries—are found in intrusions of mafic and ultramafic rock that forced their way into the Kaapvaal and Zimbabwe Cratons about 2 to 2.5 billion years ago, most likely derived from prolific melting in underlying mantle plumes. These intrusions are exposed at the surface as geological features that can be seen from space, forming the Bushveld Complex north of Johannesburg, and the Great Dyke that extends almost the full length of Zimbabwe. The mineral deposits were created when the intrusions were still molten, precipitating straight from the crystallizing magma and sinking to the bottom of the magma chamber where they accumulated as metal-rich layers. To me, this is an example of the planet offering up its internal mineral wealth directly. But for most mineral deposits, the concentration involves a much more protracted and complicated sequence of events.

Elements like gold and silver occur in igneous rocks in amounts of parts per billion. For every gram of that element, you would need to mine a few hundred tons of rock. This is clearly uneconomical. The concentration needs to be increased at least a thousandfold to make a viable mine. Nature has found a way to do this by deploying deeply penetrating groundwater that flows through cracks in the rock, giving rise to what is known as hydrothermal activity. The result is a consequence of basic physics and chemistry. At depths of several kilometers, especially if there is an active volcano or magma nearby, the water will be hot enough to react with the rocks. And, rather like a tablet of medicine in a glass of water, rock minerals dissolve. It is also under high pressure due to the massive weight of the overlying rocks, squeezing the cracks and driving up the amount of dissolved rock the water can carry. But as the hot water rises back to the surface, it cools, the pressure drops, and it can no longer hold all of its cargo of dissolved rock. Metals and their compounds start to precipitate, lining the cracks with a metalliferous veneer. Depending on what is in the water, these could be precious metals—such as gold or silver—or base metals, such as lead, zinc, tin, iron, copper, and many others. One batch of this precipitation will not achieve much. Nature, however, is far more generous. Imagine metal-rich water continually moving through the same crack over the spans of geological time. This way, thick metallic veins can grow underground. Sometimes, the cracks reach all the way to the surface, and the metals build up on the sea floor around hydrothermal vents.

The forces of erosion now get to work. Rivers wear away the bedrock, bringing the mineral veins much closer to surface where they can be reached by miners. Sometimes, the mineral veins themselves are carried away by the rivers, broken up into fragments and tossed around in the turbulent flow, eventually abandoned by the river and deposited in a layer of sediment. And yet again, ground water flowing through these sediments can remobilize and concentrate the elements. There are many variants on the themes I have described, and each deposit—usually containing a wide variety of valuable minerals—tells its own particular geological story.

GREAT EXPECTATIONS

My tour of mines in southern Africa turned out to be a wide-ranging tour of the various ways mineral deposits are created, and also the different methods of mining them. However, before going into the details for individual mines, I think it is helpful to complete the more general biography of mineral deposits I started in the previous section. This will give you an overview of how they are discovered, and what is required to both develop and operate a big mine.

It seems that ever since *Homo sapiens* first emerged in Africa, humans have shown an uncanny ability to search out mineral deposits that lie exposed at the surface, putting them to a wide variety of uses. The iron at Ngwenya in Swaziland was used thousands of years ago as a vibrant red pigment. Gold, copper and tin were discovered by African peoples long before Europeans arrived on the scene, sometimes simply as pieces of raw metal that had been eroded out of the rock and left strewn across the ground. These were fashioned into weapons, currency, and art. And in the mid-nineteenth century, European prospectors and farmers found gold and diamonds in the banks of rivers. But what is exposed at the surface is the tip of the iceberg in proportion to what lies below. The drawback is that this deeply buried mineral wealth is very expensive to reach. For every deposit that is exploited, many have been rejected as uneconomical.

Big multinational mining companies have the deep pockets needed to take the necessary financial risks. This commonly begins with buying up mineral claims made by much smaller outfits. Turning these claims into a working mine requires extensive investigation to assess the size, shape, and grade (richness) of the deposit. The first step—usually done by the minnow before it's swallowed by the big fish—is to expand on any existing chance find by looking for similar rocks in the vicinity. At this stage, the site is called a prospect. Prospectors dig long, shallow trenches, or pits, at regular intervals. From these, they collect samples of the soil or weathered rock for laboratory analysis, hoping to find traces of a rich mineral deposit. My field area in the Malolotja Nature Reserve was littered with this sort of activity, leaving strange scars on the hillsides.

On my journey through Namaqualand and Namibia, I visited a number of prospects where signs of tungsten, antimony, iron, and tin had been discovered. Tungsten forms a mineral called scheelite, which fluoresces in ultraviolet light. The best time to find it is at night with an ultraviolet lamp, looking for the distinctive fluorescence. We spent an awkward few hours bumping into each other as we followed our host geologist in the dark, while she enthused over the glowing crystals of scheelite in the walls of a recently excavated tunnel in the bedrock. She was full of optimism about their future exploitation.

The initial prospects show what is at or near the surface. The big challenge is to look much deeper, locating the mineral deposits at depth. For this, mining geologists need specialized techniques to probe the subsurface. These techniques rely on the fact that mineral deposits have distinct physical properties. They are often magnetic, and this magnetism slightly disturbs Earth's magnetic field, even if the deposit is buried beneath hundreds or thousands of meters of rock. A sensitive magnetometer can pick up these signals. Likewise, ores tend to be denser than the rocks they are hiding in, and this will alter the pull of gravity, creating an anomaly that can be identified with a gravimeter. The measurements used to be made with laborious ground surveys, but are now done in an aircraft which flies over the terrain in a series of parallel flight paths, building up a map of variations in the strength of the Earth's magnetic or gravity field.

Despite all this sophistication—certainly compared to the methods used by early prospectors—there is always an element of serendipity in a major find. The discovery in South Australia in 1975 of what is today the world's biggest copper-uranium-gold-silver mine, near the town of Roxby Downs, is a famous case. The deposit lies completely hidden beneath a sequence of younger sedimentary rocks, hundreds of meters thick, in a flat plain. The site was first identified through gravity and magnetic measurements, but when geologists tried to reach the deposit by drilling, they found that the rocks seemingly responsible for the anomaly were far too low grade to be of interest. Fortunately, the drilling engineer decided to go for a series of holes that went deeper into the low-grade rock, and the last hole finally struck the rich lode.

After the initial euphoria of a discovery like this, the mining company is faced with the huge cost of setting up the mine infrastructure and either sinking deep shafts or opening up a large pit to reach the deposit; miners refer to the part that is being mined as the ore body. All this is at the mercy of the basic economic principles of supply and demand. The mines I visited on my trip required investments of hundreds of millions to billions of dollars. In remote areas, such as Oranjemund and Rössing, in the barren Namibian deserts, roads and housing for workers must be built, along with supplies of electricity and water. Then comes the cost of running the mine. It is not just a matter of digging up the ore—it has to be refined into a pure metal or whatever form is marketable. Here, the concentration and nature of the ore become critical, dictating the chemical and physical treatment required to separate it—factors that, in turn, will determine the cost. Low-grade gold and uranium must be leached out of the host rock using a cocktail of chemicals. Refining metals such as copper, tin, lead, and zinc consumes large amounts of electricity to power the furnaces or electroplating processes. Finally, the processed ore must be transported to market. Ores that are mined in bulk, such as iron or coal, are usually freighted to the nearest port on expensive, specially-built railway lines—you may recall the line between the Ngwenya iron ore mine in Eswatini and the port of Maputo in Mozambique.

All this expenditure can make a major mine a significant part of the host country's economy. In this way, its life span becomes an important political issue, with calls to extend it far beyond the original plan. It was initially thought that the Rössing uranium mine in Namibia might last only a decade or so, but the mine is still operating over forty years later. A longer-than-planned lifespan usually comes down to the distinction, in the mining world, between provable and probable ore reserves. Provable ore reserves are those that are economically viable to mine at the time of assessment and have been rigorously quantified by drilling; these form the basis for the mine's initial business case. Probable ore reserves are more speculative and include possible future expansion of the mine. For this reason, it can be difficult to determine the true mineral wealth of a mine—or indeed, of our planet. Pessimists tend to focus on provable reserves, and optimists look at probable reserves. Whichever you choose, it is likely that we will soon start to face shortages in many of the basic ingredients of our industrial world. The global provable reserves of iron are estimated to be around 100 billion metric tons, yet we are consuming around two billion metric tons of these reserves every year. At this rate, we could run out of iron fifty years from now, unless we find significantly more reserves. The timeline for uranium is very similar.

I think we will soon be living beyond the long-term means of our planet, with demand outpacing the discovery of new resources. As Mr. Micawber said in Charles Dickens's *David Copperfield*, referring to the pre-decimal British currency of pounds, shillings, and pence: "Annual income twenty pounds, annual expenditure nineteen nineteen and six, result happiness. Annual income twenty pounds, annual expenditure twenty pounds ought and six, result misery." My father once told me, when I was thinking about studying geology, that there will always be a need for geologists who can find new mineral resources. But perhaps future geologists will be prospecting for these elsewhere in the solar system too!

- § -

GOLD

(Price in 2025: About US$110 million per metric ton of refined gold)

The first stop on our trip was one of the deep gold mines not far outside Johannesburg. In fact, no geological tour of South Africa is complete without witnessing this aspect of the country's economic life. The gold is found in conglomerates that were laid down nearly three billion years ago on the surface of the newly formed Kaapvaal Craton. It is thought that the gold originally came from veins in much older volcanic rocks, such as those in the Barberton Greenstone Belt, formed during an earlier period of mineralization. Fragments of these gold veins (as well as other minerals including small amounts of uranium and diamonds) were plucked out and carried by rivers that flowed into a wide inland plain and lake, and here they were deposited in layers of gravel. The gold may have subsequently been dissolved and reprecipitated in groundwater flowing through the gravel. Much later, these gravel layers were caught up in movements of Earth's crust, and tilted so that they were no longer horizontal. Erosion of this landscape has revealed the gold-rich layers where they reach the surface as a ribbon of rocks—called "reefs" by miners—outcropping along a prominent ridge called the Witwatersrand, usually shortened to the Rand. Their discovery by the early prospectors started the world's largest gold rush in the 1880s.

As always happens in a mineral rush, well-heeled speculators moved in after the initial finds and bought up claims. To begin with, they bought whole farms that straddled the reefs. Cecil Rhodes—who gave his name to the former Rhodesia, now Zimbabwe—was already investing in diamond mining at Kimberley, about 450 kilometers southwest of the Rand—we will get to the diamonds later in our journey—and he saw a money-making opportunity too good to miss. Unlike many of the other speculators, Rhodes and his partners realized that the gold was contained in layers that sloped into the ground— there is nothing like a bit of geological understanding to put you ahead of the game. Rather than only buying farms where the reefs reached the surface, they bought land to one side, where the reefs were deep underground. This land was cheap because few other people had associated it with gold at depth. The price Rhodes had to pay for this clever strategy was the cost of digging deep vertical shafts to reach the gold-bearing layers. It was worth it, though, because he and his company, Consolidated Gold Fields, made a fortune.

Today, the gold mining district on the Rand is dominated by mountainous spoil heaps where the waste rock has been dumped after the gold has been extracted. The original methods of separating the gold were much less efficient than modern techniques, so when the price of gold is high, there

FIGURE 6.1 The allure of gold. Clockwise from bottom: a mining geologist points to a gold-bearing conglomerate, laid down by rivers about 2.9 billion years ago and now buried several kilometers underground at the Randfontein Estates gold mine near Johannesburg; to reach it, miners are lowered down a vertical shaft beneath this massive derrick; at a small gold mine in Zimbabwe, much closer to the surface, miners use an 1880s stamping battery to crush the gold-bearing quartz veins and extract the gold. (Photo credit: Simon Lamb)

is money to be made in sifting through the older dumps, picking up the few grams of gold left by early miners in every ton of waste. The rich reefs are now mined far below the surface, but even here the rocks do not glitter with gold, and the mines are wet, noisy and dirty places. There is a certain routine for entering one of these deep gold mines. For a visit like ours, it is usually arranged for the morning. First you are taken to the changing rooms, where you put on overalls, rubber boots, and a helmet with a head-lamp. The visit will end in this same place, but with a refreshing hot shower to wash off all the grime.

The entrance to the mine is a vertical shaft a kilometer or more deep—the deepest mines are over four kilometers down. A huge derrick stands over this hole, holding the winding drums that lower and raise the lifts. The lifts are large cages, with only a wire grill separating you from the side of the shaft. Often there are separate "up" and "down" cages, so that when you descend, slowly accelerating into near-free fall, the "up" cage races past you like a freight train charging through a station, creating a roar and blast of air that can almost knock you over. As the cage approaches your destination level, it slows down. But because you are dangling on a kilometer or more of cable, you bounce up and down like a spring before finally coming to rest. The cage door clanks open, and you are admitted to a long and poorly lit tunnel with a railway track running down the middle, and deep gutters full of flowing water on either side. At these depths, mines must be continually pumped dry.

A miniature train takes you to the working part of the mine, rattling along its track as it snakes its way through the sinuous tunnels. The train stops and starts at signals and junctions, its whistle piercing your ears, and soon you lose all sense of direction. After traveling several kilometers like this in the dark, you reach a nightmarish world, lit only by the headlamps of the miners and dominated by the deafening noise of pneumatic drills and compressors, the spray of water to cool the drill bits, and the clangs and bangs of rock falling into empty wagons on the train. It is also very hot—without air-conditioning the mine would be unbearable. The mine geologists experience this every day, and they have learned to block out all the distractions and focus on their job.

At the working face—the front line of the operation—I could just make out with my headlamp a rock made up of white pebbles of quartz. Between the pebbles there were streaks and blobs with a brassy tinge. This was the ore, consisting of a mixture of gold, uranium oxide, and pyrite (iron sulfide). The brassy color is in fact the worthless pyrite, and the gold is virtually invisible—even at the high grade for gold of about two hundred grams per ton of rock, there won't be much to see. I began to recognize the distinctive wavy patterns of cross-bedding in the streaks of ore. From this, the mining geologists can work out which way the rivers that deposited the gold were flowing nearly three billion years ago. They can also recognize channels in this river where the gold is more concentrated. As they mine the bedrock, they follow these channels, sticking close to the rich ore.

Over time, the mine has become a warren of tunnels and vertical shafts extending deep underground, following the sloping layers of bedrock. The colossal weight of the overlying rock presses down on the tunnels, and the danger of collapse is ever present. The rocks are constantly cracking, making the deep gold mines of the Rand one of the most earthquake-prone regions on Earth at over a thousand earthquakes a day, though most are so small they cannot be detected without a sensitive seismometer. Occasionally, a collapse does trigger an earthquake that can be felt by the inhabitants of Johannesburg. To some this may be a reassurance that South Africa's wealth is still being mined. To others it is a reminder that gold mining can be a deadly occupation.

The ore is hauled up the access shafts to the surface and taken by conveyor belt to the mine processing plant. Here, the gold will be separated from the host rock, first by crushing the ore into a powder, and then a poisonous chemical treatment that involves large quantities of cyanide. The total annual output of the mines on the Rand was about 700 metric tons of pure gold when I visited, although it is steadily declining from a peak production of about 1,000 tons in 1970. Despite all this effort, the main use of gold in our modern society is jewelry and bullion that underpins some currencies, although it is also recognized as an exceptional electrical conductor for electronics, and for having shielding properties that protect space vehicles from solar radiation. But if you have seen native gold—gold in its elemental form—you will have experienced the strange pull of its beautiful warm color, somewhere between orange and yellow.

I once visited a small gold mine in Zimbabwe where they were using traditional techniques to burrow into the ground by hand, following a thick gold-bearing quartz vein. The gold was separated by first crushing the quartz with an antique stamping battery—a set of heavy vertical rods that pounded the rock—and then running the resulting slurry of crushed ore down a long chute lined with mats to catch the heavy gold particles. I remember saying to a colleague that you will never see the gold in the ore. To demonstrate this, I randomly picked up a piece of quartz from the heap that was about to be processed. I had to eat my words, because there, among the milky white quartz, was the unmistakable wispy, orangey-yellow seam of gold, glinting in the bright sun. I quickly pocketed it.

IRON

(Price in 2025: About US$90 per metric ton of iron ore)

The world's main supply of iron comes from sedimentary rocks originally deposited as iron-rich mud on the seafloor, known as banded iron formation because the iron is concentrated in thin layers that appear as bands in the rock face. I have already described the banded iron formation in the Barberton Greenstone Belt, made up of layers of chert interleaved with iron oxides. However, in the following billion years, far more extensive deposits accumulated at the bottom of the shallow seas that fringed the now-larger landmasses. These are preserved today in most of the ancient cores of the continents, famously in the Hamersley Range of Western Australia, and South Africa's Kaapvaal Craton. The next mine on our trip is a spectacular example of the immense scale of these sources of iron. This is the Sishen iron ore mine about 550 kilometers southwest of Johannesburg.

Sishen is a very different mining operation from the deep Rand gold mines. The mine is an open pit, about 14 kilometers long and 4 kilometers wide, coated in a fine red dust that stains everything it touches. Here, the layers of iron (in the form of an ore composed mainly of iron oxides) lie close to the surface and are only gently warped. These were originally deposited about 2.46 billion years ago in shallow water, on top of a sequence of limestones. Much later, they were bathed in near-surface groundwater, which both leached silica from the rock and reacted with the iron, enriching it to an economic ore grade. Mining the ore is a brutal operation that shatters the bedrock by blasting. The rubble is scooped up by enormous mechanical arms and dropped into waiting dump trucks. The mine works around the clock, and the dump trucks constantly shuttle back and forth between the working face and the crushing plant. When I visited, it was producing about 20 million metric tons of iron every year. During a global economic boom, similar iron ore mines around the world can barely keep up with demand, and this creates pressure to keep expanding mining operations.

Further development of the mine begins with a program of drilling to find where the iron-rich layers extend underground. Mine geologists are searching for ore that is more than 60 percent iron. Drill holes are planned with military precision to form a gridwork of dots on the map. Drilling is done by coring, in which the drill bit is actually a hollow tube with a hardened cutting edge. As this penetrates the rock, it produces a long cylinder—called a core—that is extracted from the hole. Large quantities of this core, laid out in wooden trays, were stacked on top of each other in the mine's core shed—each multicolored cylinder of rock revealed individual iron-poor and iron-rich layers as red, black, or gray bands, ranging from millimeters to centimeters (fractions of an inch to inches) thick. The mine geologists that we spoke to were confident they could identify particular patterns in the bands, appearing as variations in thickness, like a bar code on a packaged item of food, and that these patterns could be traced over tens of kilometers. Such uniform underwater conditions suggest to me that the iron-rich sediment was deposited on a virtually flat seabed adjacent to a low coastline.

Where did all this iron come from? Geologists used to think that it was eroded from a hinterland of ancient volcanic rocks such as those in the Barberton Greenstone Belt—rather like the gold and uranium found in the deep Rand mines. But there is a problem because it is unclear how so much iron reached the sea to create such a concentrated deposit. During younger periods of geological time, the iron in the detritus was quickly oxidized into red or brown minerals such as goethite and hematite—that is, it rusted—when it came into contact with oxygen-rich air. In this way, sandstones laid down by rivers are typically stained red-brown color, trapping the iron oxides. However, sandstones laid down by rivers during the Archean Eon—such as those in the Moodies Group of the Barberton Greenstone Belt (see chapter 5)—do not have this red iron oxide stain. This absence has been taken as evidence that early Earth's atmosphere was starved of oxygen, and therefore unable to oxidize the iron (see chapter 8 for further discussion). An important consequence of this is that the unoxidized iron is now soluble in water and can be transported in solution for long distances by a river, finally reaching the sea. Here, it could potentially be deposited as iron-rich layers in banded iron formation, if there were a way to oxidize the iron.

The deep ocean may have been an even more important source of iron, through hydrothermal vents on the seafloor where iron-rich water gushed out. This iron ultimately comes from the basalts that make up the oceanic crust. In oceans with very low concentrations of dissolved oxygen, rich clouds of iron could drift large distances on ocean currents, eventually reaching the shallow continental shelves. The discovery of the fossilized remains of very ancient microbes (microorganisms) has suggested that biological activity might have triggered the precipitation of the iron, thereby playing a key role in the creation of banded iron formation, as I described in chapter 4. The original idea—first suggested by Preston Cloud in 1965—was that this was due to oxygenic photosynthesis, in which cyanobacteria use sunlight to split carbon dioxide and water, making hydrocarbons. The by-product of this is free oxygen, which can then react with the dissolved iron to produce insoluble iron oxides.

But there is yet another problem here, because if widespread oxygenic photosynthesis was a source of free oxygen, why would the oceans and atmosphere be virtually devoid of it? Still sticking with a biological explanation, a possible solution is that oxygenic photosynthesis in organisms had either not yet evolved or was not widespread, and that the oxidation of iron was instead driven by a different type of microbe—photoferrotrophs. These can use light energy in a chemical reaction that causes oxidation of iron but does not produce any free oxygen. Perhaps massive precipitation events were triggered by blooms of these microbes, controlled by the seasons. The heavy iron particles, in the form of oxides and hydroxides, would then settle out on the shallow seafloor, building up iron-rich layers that alternated with iron-poor, silica-rich mud deposited during periods when the microbes were less abundant.

Whatever the source of the iron—on land or underwater—it is clear that these deposits are evidence for a time in Earth's history when there was virtually no oxygen in the atmosphere or deep oceans. I find it an intriguing thought that our modern civilization, with its insatiable demand for iron, could only develop because there was a time on our planet when we would have been unable to breathe.

COPPER

(Price in 2025: About US$9,500 per metric ton of copper metal)

After the dusty mine at Sishen, it was a relief to be on the move again, as we headed farther southwest into Namaqualand. We were travelling to the copper mining district known as OKiep (or confusingly, O'okiep), whose mining history dates back nearly to the beginning of European colonization of South Africa in the mid-1600s. Here, we encountered a geology quite different from that of the previous mines—we were entering a volcanic realm. Erosion has laid bare the deep roots of volcanoes, consisting of irregular pipes and dikes of volcanic rock akin to basalt—called norite—which had forced its way up into the Namaqualand basement about one billion years ago. It is these that are being mined for copper.

So, what is the origin of all this copper? One idea is that it comes from oceanic crust that got caught up in the Namaqualand basement during an episode of continental collision, and was later brought to the surface when this crust melted to form norite magmas. It is well known that the rocks beneath the seafloor are riddled with veins of copper, scavenged from the crust by deeply circulating sea water and then precipitated in cracks—this is the same source as some of the iron in banded iron formations—eventually gushing out of hydrothermal vents. For this reason, fragments of oceanic crust that have been pushed up on land—called ophiolites by geologists (see chapters 1 and 3)—have proved to be rich sources of copper. During the Bronze Age, about 3,200 years ago, large quantities of this metal were extracted from the Troodos ophiolite on the island of Cyprus, giving us the word copper from a corruption of the Latin phrase *cyprium aes* meaning "Cyprus metal."

The copper at OKiep occurs as copper sulfide which separated directly from the magma. This was subsequently enriched by the movement of ground water near the water table, creating narrow ore bodies that extend several hundred meters beneath the surface. This means that mining is an underground operation and involves digging shafts and tunnels to reach the copper. The particular mine we were visiting had a rather unusual access point. The miners had built a spiral shaft that corkscrewed down into the ground, circling the norite body. You actually drove down this spiraling ramp in a jeep, although this was a hair-raising experience. It had been constructed with the tightest curves that would still allow access without having to put the vehicle into reverse. At full lock, our jeep continuously grazed the outer rock wall of the spiral, but the driver never slowed down.

The mining company had discovered that the best way to mine the copper is to just remove the norite wholesale, leaving the surrounding rock behind. To do this, they drill into it from all angles, until the norite body is riddled with holes. The holes are filled with explosives, so that when the rock is blasted, it shatters into a loose rubble. The rubble is removed from the bottom of the mine, relying on gravity—like sand flowing through an hourglass. As this happens, an empty space is left behind. This is called open stope mining, and it has created huge underground caverns. I remember being worried about the danger of a catastrophic collapse of the roof, creating a deep sink hole. Fortunately, that has not happened yet, and nobody lives on top of the mine anyway.

After the morning trip underground, and a shower in the changing room, we were treated to mine hospitality. That day, it was a barbeque—or braai, as they say in South Africa—on the grounds of the mine's social club. We were provided with braai packs: piles of steak and coils of boerewors—a fatty sausage—which we took to the grill to be cooked. And there was plenty of beer. Yet the mine geologists were already wondering how long this lifestyle would continue. The price of copper was falling rapidly, having lost nearly half its value in the previous two years, and there were many rich copper deposits in the region that were being abandoned.

DIAMONDS

(Price in 2025: About US$30 billion per metric ton of natural 1-carat cut diamonds)

We now drove due north to the Orange River, and on into Namibia. At that time, Namibia was not considered a separate country by South Africa, but rather a province called South West Africa, or SWA for short, with administrative rights granted after the First World War. The southern border of

SWA was the Orange River, which flows all the way from the Drakensberg Mountains in Lesotho, on the eastern side of southern Africa, with a catchment that includes most of the Kaapvaal Craton.

The Kaapvaal Craton hides, in its deep keel over 150 kilometers beneath the surface, an incredible wealth of diamonds. The diamonds are made of carbon atoms—the same stuff that forms coal—crushed into diamond by the huge pressures at these depths. But there is a way for them to get to the surface. Occasional volcanic upwellings from much deeper in Earth are full of explosive volatiles such as carbon dioxide and water. Called diatremes, these have the power to punch their way right through the craton. Their remains can be found in many parts of Africa—they are particularly numerous in southern Africa—forming pipe-like bodies of strange rocks from Earth's mantle. And some of them are rich in diamonds that have been plucked from the bottom of the craton. The first major diamond fields discovered in South Africa—in 1869, around what is today the town of Kimberley—turned out to be the remains of several diatremes. For this reason, diatremes are sometimes referred to as kimberlite pipes.

The Kimberley diatremes erupted 78 to 92 million years ago, in the Cretaceous Period, and the debris from the eruptions was peppered with gem-quality diamonds. The diamonds themselves are much older, crystallizing deep in the mantle during the Archean Eon, around 2.9 billion years ago. In the mad scramble to mine them, the miners dug down ever deeper—one of the mines became a circular open pit, about 460 meters across and over 200 meters deep, known today as "the big hole of Kimberley." Cecil Rhodes made his first fortune here in the 1880s, buying up the rabbit warren of small claims in and around this hole. Rider Haggard also had this hole in mind when describing, in his novel, the source of King Solomon's diamonds—a deep, circular pit with steep sides, at the bottom of which the heroes eventually emerged through a fortuitous animal burrow that connected to the maze of underground rocky tunnels where they had been trapped.

The Orange River has had time enough to spread its tentacles over much of the Kaapvaal Craton, seeking out hidden diatremes and the diamonds they contain. The diamonds have been carried downstream, invisible among all the sand and mud, but there all the same. Some of these diamonds have become stranded in gravels deposited on the banks of the river. But many of them have made it to the Atlantic, and then northward along the coast in the longshore drift, tossed around on the watery conveyor belt—it is mainly the unflawed stones that survive the long journey. Finally, they have come to rest on the sea floor, or stranded on ancient beaches, trapped in small depressions or holes in the bedrock.

Over time, the Atlantic coast of southern Africa has slowly risen relative to sea level. The reasons for this are still debated, but there are very practical consequences. Pockets of diamond-rich gravel that were once on the sea floor now lie onshore, in the extremely arid Namib Desert. This, however, is not the end of the story. Strong winds whip up the desert sands, sculpting them into wave after wave of beautiful yellow dunes that can be hundreds of meters high. Blown by the wind, the dunes move over the pockets of diamonds, covering them with vast quantities of sand in a region that reaches inland for about 100 kilometers and extends north of the mouth of the Orange River for about 320 kilometers. It is also the world's largest mine. On some maps it is marked as the "Gebiet Verboten," or Forbidden Zone, although it is more commonly known as the "Sperrgebiet," or Restricted Area. Whatever you call it, without permission, you enter at your peril.

The mine—called Oranjemund, Afrikaans for "mouth of the Orange River"—was run by De Beers, and, when we visited, its output was several hundred kilograms of diamonds each year. We were met at the gate into the restricted area, having driven nearly 100 kilometers through the desert along a road where you were not allowed to stop. Our hosts were a group of mine geologists working for De Beers. We left our vehicles and transferred to theirs. No vehicle was allowed to leave the mine working area, and so the mine had its own fleet of vehicles. When they broke down or got too old, they were driven to a graveyard of used vehicles. We drove past it, gazing at row upon row of jeeps, trucks, diggers, earthmovers, and cranes, all left to rot in the desert. One began to appreciate the scale of the mine after being driven for an hour along sandy tracks without seeing any sign of activity. Finally, we reached the center of the operation.

Mining diamonds requires an extraordinary range of techniques. First, the overlying sand has to be removed. This is done with teams of giant

FIGURE 6.2 Digging up more of southern Africa's mineral resources. Clockwise from bottom left: the Big Hole at Ngwenya, at the southern end of my study area in the Malolotja Nature Reserve, Eswatini, left behind after the mining of high-grade iron ore from Fig Tree Group banded iron formation and ferruginous cherts and shales; another major mining operation—this time for diamonds at Oranjemund on the Atlantic coast of southern Namibia—involving the bulldozing of massive sand dunes that cover diamond-rich beach gravels; and a skilled worker at Oranjemund sorting through concentrated diamond-bearing gravel, picking out individual jewel-grade diamonds. (Photo credit: Simon Lamb)

earthmovers that scrape up the sand and dump it elsewhere. The sand, however, is so soft that they soon get bogged down. The miners have anticipated this, and they have a giant bulldozer waiting. It comes to the rescue by ramming the back of the earthmover with its blade. This combination struggles on for a few tens of meters before getting stuck again. But there is another bulldozer ready, which rams the back of the first bulldozer, and the train of heavy equipment moves forward a little bit more. All this is repeated until there is enough brute force to push the earthmover through the sand and out to where it can dump it. The operation will continue for weeks or months until the giant sand dunes have been completely cleared away.

Now comes the next stage, and this is more delicate. Mechanical diggers clean up the underlying bedrock until small pockets of gravel, a few meters across, are revealed. Each pocket is then excavated individually by miners with pneumatic drills. Finally, a team of miners descend on the bare rock with dustpans and brushes, and they sweep up the last bits of gravel and sand at the bottom of each pit—this is where the richest deposits of diamonds lie hidden. We were told that the mine is beginning to shift its efforts to underwater mining. This is done by sucking up the sand and gravel pockets in the shallow water, just offshore of the present operation, in a giant suction device. The equipment is on boats or moveable platforms that systematically work the seabed. Environmentalists, however, are concerned that this could have an impact on marine life, because it causes local disruption of underwater ecosystems.

All the excavated debris is taken by conveyor belt to the separation plant. Here, it is washed and shaken to concentrate the diamonds—the diamonds are relatively heavy, so they sink to the bottom. The final product is a diamond-rich gravel spread out on a conveyor belt that slowly passes teams of eagle-eyed diamond pickers. These people intently watch the moving ribbon of gravel, expertly picking out the diamonds as they pass by. It was fascinating to stand in the separation shed and see this happen, watching a small pile of gem-quality diamonds steadily accumulate next to each picker. The diamonds are of all sizes, ranging from a sand grain to a large piece of grit. I was surprised to see that they are not always transparent like glass—a few are pale blue, yellow, or green, depending on the impurities in them. Some colors are very rare, and these diamonds can be the most valuable. But we were told that De Beers mainly keeps them for their diamond museum. At the time of our visit, the total output of the mine was a significant percentage of global demand, and De Beers held back diamonds to keep the price high. Mines like this demonstrate that there is no shortage of diamonds and that their rarity value has been greatly inflated by clever marketing. To me, they are far more interesting from the point of view of what they tell us about the conditions deep inside our planet—they are some of the deepest samples we have.

Leaving the diamond mine was as intimidating as entering it. The big problem for De Beers was that it is very easy to steal diamonds. If you are one of the workers brushing up the sweepings in the rocky pits, or a picker on the conveyor belt, it is not hard to pocket a diamond. The trick is to swallow it. De Beers tried to prevent this by carefully monitoring the workers with CCTV surveillance. The company's safety net was a policy of not allowing anything but people to leave the mine, so diamonds could not be stashed away. Everybody had to pass through an x-ray machine—diamonds are opaque to x-rays and so easily spotted in the body. We passed through these machines ourselves as we left the mine, and I remember feeling nervous that I might have somehow picked up a diamond by mistake. After all, there is a chance that some grains of sand in the dunes are diamonds, blown there by the wind.

URANIUM

(Price in 2025: About US$170,000 per metric ton of uranium oxide, known as yellowcake

After Oranjemund, we continued north toward Windhoek, the capital of Namibia. We were making our way to a uranium mine called Rössing, operated by Rio Tinto Corporation. The route took us near the Kuiseb River. Here, the journey was enlivened by a curious anecdote told to me by the leader of our trip, which I think illustrates the dedication of geologists to their subject and also the remoteness of the region. During the Second World War, German nationals in southern Africa were interned because they were considered a potential security threat. Namibia was a German colony before the First World War and still has a sizeable population of German-speaking settlers today. Two German geologists—Hermann Korn and Henno Martin—slipped away and camped in rock shelters hidden in the Kuiseb River valley, avoiding detection for two years. They used their time well, taking the opportunity to make the first detailed geological map of the unexplored rugged terrain. Later in the war, they emerged again, and their map was eventually published, generating considerable interest among the South African geological community. Their wartime experiences in the Kuiseb River Valley completely occupied my mind as we drove on through the Namib Desert, because I was trying to envisage the day-to-day realities of surviving for such a long time in this dry and rocky landscape while avoiding contact with locals who might betray them.

When we finally reached our destination at Rössing, it was quite a shock to be confronted with the huge infrastructure of a large mining operation. The mine was an open pit, and although it had only been in operation for about six years, it was already one of the world's largest producers of uranium. The miners were chasing gashes in the bedrock that formed veins filled with large crystals of quartz and feldspar, some measuring several centimeters across. The veins were a whitish-gray with glassy surfaces that reflected the strong Namibian sunlight, producing a blinding glare that hurt my eyes. We were told by the company geologists that these were the last gasps in the cooling and crystallization of a magma called alaskite (presumably, a similar rock had been first discovered in Alaska!) which had formed during volcanic activity about 500 million years ago. And some of the veins contained hairline cracks filled with a yellow uranium mineral—this was the uranium ore that made the mine commercially viable. The uranium was thought to originate deep in the crust, where it had been concentrated as part of the process that made the continental crust in the first place.

The mining operation was working around the clock to make enough profit to recover the initial investment. The dire political situation in Namibia, with an ongoing guerrilla war on the Angolan border, put Rössing at the highest risk level in Rio Tinto's assessment. In fact, Rössing is still working today, over forty years after my visit. The current pit extends over an area of about 4.5 square kilometers and is nearly 400 meters deep. Mining here is the usual rock blasting, followed by loading the shattered rock into giant dumper trucks that take it to the processing plant. What is unusual, though, is that the ore is very difficult to see. However, it is radioactive, as uranium spontaneously decays into thorium and lead (as well as radium, radon and other daughter products), emitting alpha and beta particles, and gamma radiation. So, the only way to decide whether to process or dump the rock is by measuring its radioactivity. The dumper trucks drive through a shed rigged with Geiger counters. If the reading is above a certain level, the dumper truck goes to the processing plant. Otherwise, the load is dumped in an old part of the mine.

I was to learn more about the radioactive levels in the ore when I collected my own sample of alaskite from the mine. This was an angular lump of quartz and feldspar, about the size of my fist. I put it in my pack and brought it back to Cambridge. One of my colleagues had a Geiger counter. This had both a visual meter and an audio output that normally emitted the occasional clicking sound. But when we put it near my sample, the audio output became a steady purr. Word soon got around the department that I had a radioactive lump of rock. Stuart Agrell, in charge of the department's mineral collection—the Harker Collection—came to see me and immediately confiscated my sample. My understanding is that it is sealed in a lead box as a prized part of the Harker Collection. To think that this rock had been in my luggage, and even in my pocket, for several months!

The radioactivity changes the nature of the rock itself, because the emissions rip through the crystals and leave behind a trail of damaged rock. Quartz loses its milky-white color and becomes a smoky gray. The damaged crystals are also exceptionally hard. The geologists at Rössing told us that the alaskite was the hardest rock they had ever encountered in a large-scale mining operation. The sharp pieces of rock gouge out the steel in bulldozer blades or digger buckets, and rip the rubber conveyor belts to shreds in a matter of days. The mine engineers have come up with a very simple solution. They weld extra-thick steel plates to the blades and buckets, and double or triple the layers of rubber on the conveyor belts. Even this does not last long, adding to the cost of the mining operation.

Processing the ore requires large quantities of water, a cause for concern because water is a scarce commodity in the Namib Desert. The water is used to wash the crushed ore in a toxic soup of chemicals that leach the uranium oxide from the rock. Only a small fraction—much less than 0.1 percent of the mined rock—is uranium oxide. The final output is a powder called yellowcake, and the mine produces a few thousand tons of this each year. After further enrichment, which increases the proportion of radioactive uranium-235 relative to uranium-238, the material can be used as fuel rods in a nuclear power station.

Rössing was the last major mine we visited, and we reached our farthest point of the trip soon after, at the old German settlement of Swakopmund on the Atlantic coast, complete with its pier and thick sea fog. This is where I first heard about the outcome of the Falklands War. It was June 1982, and I was drinking beer with other members of our group in the bar of our hotel. The radio was blaring on the loudspeaker system, the voices mostly in Afrikaans. Then I suddenly heard a news flash in English announcing that British soldiers had captured Port Stanley and the Argentinian Army had surrendered. I had been on tenterhooks about the outcome of the war for weeks; the task force was still sailing to the Falklands when I flew out to South Africa. Now it was all over. The response in South Africa was very mixed, as there was ambivalence about which side to support. I just felt relieved. Anyhow, the end of the war was also the end of my tour of southern Africa's mineral wealth. I returned to Swaziland so that I could continue where I had left off with my own field work.

- § -

AN UNUSUAL TELEPHONE CALL

The experiences I had on my trip in Namaqualand and Namibia were much more than visiting mines. The following anecdote illustrates the quirky side of South Africa in the early 1980s. We were passing through the small township of Pofadder on our way to the OKiep copper mining district. Pofadder sticks in my mind for two reasons. First, I was puzzled by the name—perhaps a place where these snakes were common (in case you have not guessed, pofadder is Afrikaans for puff adder)? Why would anyone put a town here? And second, this is where I made an unusual telephone call. At that time, many parts of South Africa's rural telephone network did not have direct lines. A telephone call involved first ringing the operator, who then made the connection with the number you were calling. I tried doing this from the telephone box in Pofadder.

I wanted to ask British Airways in Johannesburg to change the date of my flight home. The telephone had strict instructions: pick up the receiver and wind the handle on the side of the phone only once, then wait. Winding the handle rang a bell to alert the operator that somebody wanted to make a call; I could see how its overuse would become irritating. A bad-tempered voice finally answered in Afrikaans, but after such a long pause that it was only with great restraint that I had not wound the handle again. It then became a battle of wills: I spoke in English; the operator, Afrikaans. I was sure the operator had perfect English, and I certainly knew no Afrikaans. I gave her the number of British Airways, and then put the receiver down and waited. Eventually, my phone rang again and I could hear the operator speaking to British Airways—in English this time—explaining that there was a call from Pofadder. The line remained open while the operator told me how many coins to put in the coin box—it was eleven, though I can't recall which ones.

Of course, I did not have enough change. I asked the operator to wait while I had an urgent conversation with a companion waiting outside. When I had cobbled together what I thought was enough, I dropped them in the box. The operator was counting the clicks, which she could hear on her line, to make sure that I was not cheating. She claimed I had only used ten coins, while I was convinced it was eleven. We had a protracted argument of the yes-I-did, no-you-didn't variety. I was certainly not going to concede: I had no more coins. Finally, with great reluctance, she let me talk directly with the British Airways agent. All this had taken nearly ten minutes on the open line, and it must have made quite an impression at British Airways, because when I checked in at Jan Smuts Airport (now O. R. Tambo International Airport) in Johannesburg a couple of months later, the agent at the check-in counter remembered my call. I suspect it had entertained British Airways staff for weeks afterwards. After all, it is not every day that you get a telephone call from Pofadder.

Act III

7 | The Face of the Earth

How Earth got its essential features

We are approaching the end of our quest to discover what our planet was like when it was still young. We have focused in the previous chapters on the landscapes and seascapes that existed 3.5 to 3.2 billion years ago, revealed by the rocks of the Barberton Greenstone Belt and exposed in southern Africa's Makhonjwa Mountains. Reading these rocks has not been easy. It has required extensive fieldwork and laboratory investigations, together with a wide knowledge of physics, chemistry, and biology. Another breakthrough has come from using our observations of how similar rocks are forming today, such as those on the seafloor or at the edges of the tectonic plates, or deep inside Earth. In this way, it has been possible to link what we see in the ancient rocks with both modern environments and the deeper workings of our planet. All this research has allowed us to visualize our planet in surprising detail, when it was about a quarter of the way through its long history.

OCEANIA

We can now answer the question I asked near the beginning of this book: how would the surface of the Earth have appeared to alien explorers (if such things existed and they saw reality the way we do) traveling through the solar system billions of years ago, during the early years of our planet's story? My conclusion is that they would have found a place, in many respects, very familiar to us.

The rocks of the Barberton Greenstone Belt clearly show us that it was a world with extensive oceans and intense volcanic activity on the deep seafloor. Here, magma continually spilled out of cracks, rapidly chilling in the surrounding water as pillow lavas and creating the sort of rugged undersea volcanic landscape that we find today at seafloor spreading centers. Superheated water gushed out of vents, carrying minerals that precipitated around them, building up chimneys of rock—just like those commonly seen in deep sea dives on submarine volcanoes. And, like today, life was thriving around these vents.

Volcanic islands rose up from the ocean depths, either fed by magma from upwelling mantle plumes or forming volcanic arcs above subduction zones. Like their modern counterparts, these islands were dangerous places: pools of hot bubbling mud dotted their shores, and clouds of volcanic ash periodically exploded from volcanic craters. Life was here, too—forming microbial mats that covered the seabed in the sheltered nearshore waters, or building up distinctive mounds of limestone in the intertidal zone. Periodically, large earthquakes violently shook the seabed, triggering submarine debris avalanches that cascaded down into the deep ocean, creating vast jumbles of rock similar to those that have been found on the steep edges of ocean trenches.

All this geological activity was transforming the early world of oceans into one with small continental landmasses—still rimmed by vast seas. At the edges of this land, the ocean waves moved back and forth on sandy beaches, forming a coastline punctuated by bays, lagoons, inlets, and estuaries. The tidal range was similar to today's, scouring the seabed in more open water; and if we had looked up at the night sky, we would have seen our familiar moon. Large rivers brought muddy water from the continental interior, and the mud settled on the seabed. Inland, these rivers flowed across wide flood plains that had been built up with sandy and gravelly deposits. And farther in the distance, the headwaters of these rivers drained a mountainous terrain, perhaps enveloped in thick cloud—I think these mountains were pushed up just like much younger mountain ranges, caught between two converging tectonic plates, first in a subduction zone and then during continental collision.

Looking down on early Earth from space, I think we would have seen a blue planet, because, like today, extensive oceans of water would have scattered light in the blue part of the color spectrum. Atmospheric nitrogen would have given the sky similar scattering properties, so it too could have appeared blue, although some geologists have argued for a haze of methane, or possibly sulfur dioxide, adding a yellowish or even pink, smog-like cast (see chapter 8 for more on the composition of the early atmosphere). And if early life forms were not green, then they would certainly have been some other bright color—pink and/or purple have been proposed—adding more color, along with the reddish-brown of iron, to the shallow parts of the oceans or to lakes on land.

If I had to pick a part of the world today that best helps us see how all these images of early Earth fit together, I would choose a corner of Oceania in the southwestern Pacific, extending north from New Zealand to Fiji. This is mainly a world of oceans—though with small fragments of continents, too—rocked by great earthquakes where the Pacific Plate is sinking back into Earth's interior along the oceanic trench stretching between New Zealand and Tonga, or where the two sides of the continent of Zealandia face each other head to head in continental collision. The region is also subject to extremely explosive volcanic eruptions from a chain of volcanic islands that follow the deep oceanic trench. West of this arc, there is intense volcanic activity on the seafloor at spreading centers that lie directly above the subducted Pacific Plate. Some of the lavas here—the so-called boninites—are the closest modern examples to those strange komatiites that are so common in the Barberton Greenstone Belt. So, it seems that, in Oceania, we can witness many aspects of the workings of early Earth in action.

Here, I feel as though I have something in common with a biographer attempting to write the life of a child prodigy. In the case of the composer Mozart, it seems that by the age of five he was able to write fully formed

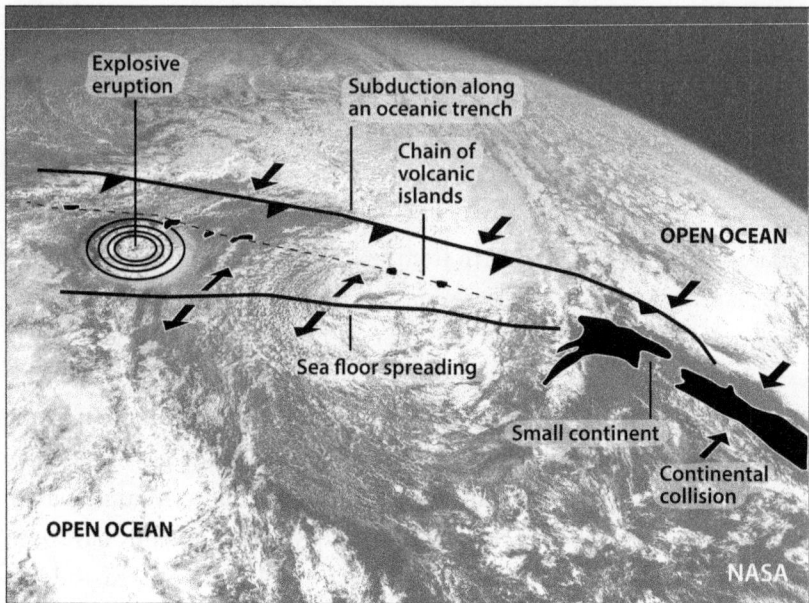

FIGURE 7.1 Plate tectonics on a world of oceans. The southwest Pacific may be our closest modern example of what Earth's surface looked like around 3.2 to 3.5 billion years ago, when the rocks of the Barberton Greenstone Belt were being laid down. In Oceania today, small active volcanic islands and continental fragments lie along the boundary between two tectonic plates that are converging—via subduction to the north and continental collision in South Island, New Zealand. Seafloor spreading occurs behind the trench, where some volcanoes erupt a distinctive, magnesium-rich lava called boninite—the closest modern analog to the komatiitic basalts that were so common on early Earth. In 2022, a violent eruption of the Hunga volcano burst from the ocean, creating vast clouds of volcanic ash that eventually rained down and settled on the seafloor and land. Eruptions like this generate vast numbers of lightning strikes, and this may have helped create the basic chemistry of life on early Earth—nurtured later in volcanic craters.
(Satellite image credit: NASA; illustration credit: Simon Lamb)

pieces of music that have stood the test of time. But this leaves very little time for what must have been a major leap in musical development. I think Earth shows an analogous trait. What we take for granted on our planet in terms of the fundamental nature of its surface is something that was created very quickly, although this has changed significantly in its details. To truly understand this, we need to complete our quest by traveling back in time to the beginning—to year zero on Earth. From there, we can retrace our steps and look for evidence of how the world we know today fully came into being. We can do this with much greater confidence now that the rocks in the Barberton Greenstone Belt have given us a well-established waypoint in our journeys back and forth through time.

IN THE BEGINNING

The birth of Earth is part of the birth of the solar system, and this is tied up with the origin of the sun. There are two fundamental observations that help us understand all this. First, the sun is a star, so we are talking about the origin of stars. Second, the planets move round the sun with orbits that all lie approximately in the same plane. These observations can be put together with the extraordinary color images of the Orion Nebula, obtained by the Hubble Space Telescope, and now the new James Webb Space Telescope. The images clearly show the events that must have led to the creation of our own sun and its orbiting planets. It all starts in a so-called stellar nursery, created by the gravitational collapse of an interstellar cloud of gas and dust. This can be triggered by the shockwaves from the explosion of a neighboring star—such events are known as supernovae. They occur when the core of a massive, dying star collapses inwards, liberating vast amounts of energy that ultimately causes an explosion like a giant atomic bomb. In fact, over time, supernovae must feed interstellar clouds with new elements created in their violent explosions.

Once collapse of the interstellar cloud has been triggered, it rapidly shrinks into a central ball consisting mainly of hydrogen and helium gas. When this becomes dense enough it will undergo nuclear fusion, igniting as a hot and glowing sun. This is how our sun must have been born. But the dust that remains farther out in space forms the solar nebula—this is the stuff from which the planets in the solar system are made. The solar nebula acquires some of the angular momentum of the galaxy, turning into a swirling disc orbiting the sun and clumping together in the inner part as rocky planetary bodies—the Orion Nebula has many examples of rotating protoplanetary disks like this. It has been calculated that within a few million years the embryo planets will have grown to sizes somewhere between those of the moon and Mars as they accrete more and more of the solar nebula. The disk will eventually become the orbital planes of the planets in the solar system.

Some remnants of the solar nebula are still in space, compacted or fused into rocky bodies such as asteroids. Every so often, one of these collides with Earth, forming meteorite showers. Avid meteorite hunters have assembled large collections that reveal many different varieties of rock. Some are just iron and nickel—these are the iron meteorites—whereas others, called chondrites, have compositions closer to that of mantle rocks. The name "chondrite" comes from the fact that the meteorite is made up of spherical clusters of minerals called chondrules. One class of chondrite, the carbonaceous chondrites, is particularly interesting because these meteorites contain water and carbon compounds, such as amino acids—potential building blocks of life. They have also proved to be the oldest known fragments of the solar system, dated using the radioactive decay of uranium to lead.

The precision of these ages is quite staggering. The oldest age measured so far is 4.56822 billion years, with an uncertainty of plus or minus 0.17 million years, for a calcium- and aluminum-rich part of a carbonaceous chondrite found in the northwestern Sahara Desert. Put in terms of human ages, this uncertainty is equivalent to one day in the life of an eighty-year-old, or a relative precision of 40 parts per million. The abundances of elements in these meteorites—except for the so-called highly volatile elements that readily form gases in the atmosphere, such as hydrogen and helium—are very similar to those in the sun. The reason we know this is because incandescent elements in the sun emit a characteristic radiation that can be used to identify them. Here is the evidence that these meteorites are indeed relics of the solar nebula, and their age is our best estimate of the birth of the stuff that made up embryonic Earth. And we can date this with a greater relative precision than any other known geological event!

I was once shown a piece of the famous Allende meteorite which fell as a meteorite shower over Mexico in 1969. In all, about 2,000 kilograms of rock were recovered, and they proved to be from one of the very ancient carbonaceous chondrites. A fragment of this meteorite was in the possession of Stephen Moorbath, who had pioneered the dating of the planet's oldest rocks—I wrote about Stephen in chapter 2. I overlapped with him during my time at Oxford University, in the Department of Earth Sciences, and I took over his office when he retired. Sitting in a small cardboard box on the windowsill of my new office, the meteorite looked like a rather ordinary dark piece of rock, with smooth faceted sides. It was hard to believe that this was a sample of the oldest known objects to be found on Earth, crystallizing at the creation of the solar system.

A GIANT PEACH

In my first chapter, I likened the Earth, with its thin skin of crust, to a giant stone fruit. A peach makes a good analogy. In the deep interior of our planet is the core, like the stone at the heart of a peach. Overlying this is the

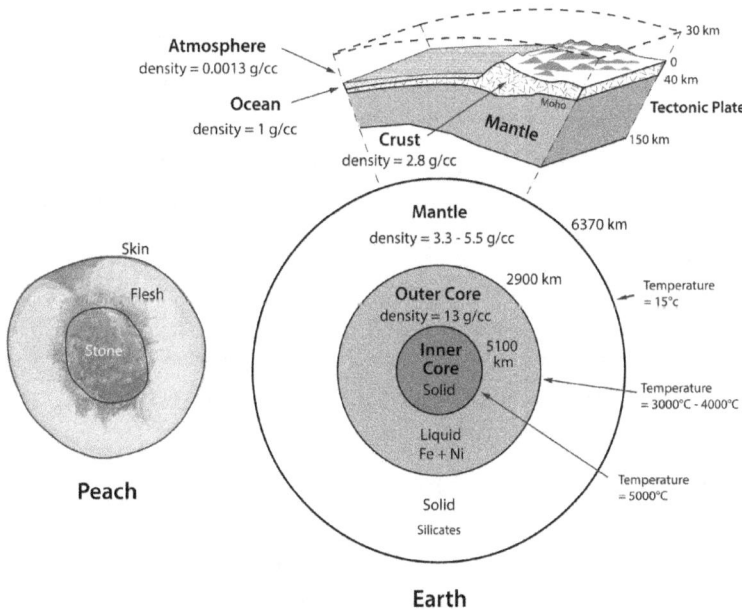

FIGURE 7.2 Our planet as a giant stone fruit. Earth's interior consists of distinct layers at progressively lower temperatures towards the surface. At its center is a very dense iron-nickel core, like the stone in a fruit, overlain by less dense silicate rocks in the mantle. The colder top of the mantle forms the tectonic plates, which include the light, silica-rich crust. A covering of water and a gaseous atmosphere make up the outermost layers of the planet. (Illustration credit: Simon Lamb)

mantle, like the juicy flesh of the peach, which is itself sealed by a thin outer skin of crust. And a covering of ocean and atmosphere equates to the fine downy hair on the peach skin. The Earth, however, would not necessarily have started out with all these layers when it was first created. Enormous amounts of energy must have been released in the accreting solar nebula during the violent collisions of rock fragments, and this would have generated enough heat to melt rock. So the starting point for the Earth was, at least in part, a molten planet. As I have described in previous chapters, the formation of the crust has taken place over geological time. The big question is: when did the other layers form?

Geochronologists have had to be very cunning to tease out this timetable. Remember that the same element can have different isotopes, some of which are radioactive. Despite showing almost identical behavior in chemical reactions, they have slightly different atomic masses. Some radioactive isotopes no longer exist on Earth because they have all decayed away, and all that is left is the daughter product. These short-lived radioactive isotopes were in the original solar nebula, most likely newly minted in the supernova that triggered the creation of the solar system. As I will explain, by measuring the abundance of their decay products it is possible to show that everything was most likely in place—core, mantle, some sort of crust, atmosphere (and by inference, oceans), and even the moon—within 100 million years of the birth of the solar system, or roughly the first 2 percent of Earth's history. In other words, if we equate the age of Earth to that of an eighty-year-old woman, the main steps in her development had already occurred by the time she was a toddler. This means that by about 4.45 billion years ago, Earth already had its essential features.

The formation of the core was an important event in Earth's history, because it gave rise to the magnetic field, and it is widely thought that without this field, life would not have survived on the planet. The magnetic field interacts with cosmic rays and the solar wind, creating a shield in space that has helped to preserve the atmosphere by reducing the ability of the solar wind to strip it away. The discovery of the core's existence, however, has involved considerable detective work. We know it is a body the size of Mars that sits right at the center of our planet because vibrations created during earthquakes bounce off it, and these can be routinely detected via a global network of seismometers. We also know from the behavior of these seismic waves that the outer part of the core is liquid, unlike the rest of Earth's interior. This liquid slows down certain vibrations (P waves) and completely blocks others (S waves), creating distinctive shadow zones on the globe for earthquake vibrations. The composition of the core is thought to consist mainly of iron and nickel; this is the only way we can explain both the magnetic field—generated by the dynamo action of swirling liquid metal—and the planet's vast mass, given the abundance of iron and nickel in the solar nebula. This composition is confirmed by iron meteorites, which are iron-nickel alloys and, almost certainly, fragments of core material from other parts of the solar system.

The huge mass of iron and nickel in the core would have separated out from the rest of Earth under the force of gravity, pulled towards the center of the planet. But when did this happen? This is where the decay of short-lived radioactive isotopes with strange names comes in. One of the pioneers of this dating technique is Alex Halliday, who was a professor of geochemistry at Oxford University when I was a lecturer in the geology department— thanks to him, I have a firsthand explanation of how it works. Alex published a series of papers in the 1990s about the age of Earth's core (and also the moon) based on studies of the abundance of tungsten and hafnium on our planet. He was making use of the fact that there is a radioactive isotope of hafnium (in addition to several non-radioactive ones), called hafnium-182 (hafnium with an atomic mass of 182), that does not occur naturally on Earth. However, hafnium-182 can be created in a nuclear reactor, and so we know that its half-life is about 9 million years. The reason radioactive hafnium-182 does not occur naturally is because with such a short half-life, there has been enough time for it to all decay away—in fact, within the first few tens of millions of years after the birth of the solar system, almost all of it had gone. Its decay product, however, does occur naturally. This is tungsten-182 (that is, tungsten with an atomic mass of 182).

Alex realized that hafnium and tungsten have another useful feature: they have very different chemical properties. Hafnium is what is called a lithophile (from the Ancient Greek *lithos*, meaning "rock," and *philos*, meaning "love")—rock lover. Its decay product tungsten is what is known as a siderophile (from the Ancient Greek *sideros*, meaning "iron")—iron lover.

What all this means is that, given a chance, tungsten will tend to bond with iron, whereas hafnium is mainly found in iron-poor rocks such as those in the crust. This is highly significant when considering the formation of the iron-rich core, because tungsten will have a much greater tendency than hafnium to follow the iron.

Geochemists assess the abundance of an isotope by comparing it to one of its stable versions that is not part of radioactive decay—in other words, an isotope that is neither radioactive itself nor the decay product of another radioactive isotope. For tungsten, this is tungsten-183 or tungsten-184. And they use the abundance of the isotope in a carbonaceous chondrite as a reference, because this is assumed to most closely represent the original solar nebula. The argument requires a clear head to follow (don't worry if you find yourself confused), but in essence, it is just a matter of using the short half-life of radioactive hafnium-182 as a yardstick to measure how long after the birth of the solar system Earth's core separated out from the rest of the planet. If this separation occurred very late after its birth, then virtually all hafnium-182 would have already decayed away into tungsten-182. This would effectively make the measured abundance of tungsten-182 on our planet the same as that in a carbonaceous chondrite, because these chondrites were never subjected to any separation of iron and nickel. But if the core formed almost immediately after the birth of the solar system, then the tungsten taken into the core would be mainly stable versions of tungsten because there would have been very little time to generate tungsten-182 through the decay of hafnium-182. However, the hafnium-182 left behind would subsequently decay to tungsten-182, resulting in a slightly higher abundance of this isotope in tungsten today compared to that in a carbonaceous chondrite.

It turns out that the measured abundance of tungsten-182 today *is* slightly higher than in a carbonaceous chondrite, and so the core must have formed before all the hafnium-182 had decayed away. In fact, we can pin this down to tens of million years, rather than hundreds of millions of years, after the birth of the solar system. As we shall see, this timetable must be similar to that for the formation of the moon. The existence of Earth's core very early on in the planet's history tells us that the dynamo for the magnetic field could have been functioning almost from the beginning. This may have been an important factor in nurturing the early atmosphere and the emergence of life.

THE MOON

One of the great scientific achievements of the Apollo missions was to take scientific instruments to the moon and bring back samples of lunar rocks. Seismometers measured the speed of moonquake vibrations through the moon's deep interior. This, together with measurements of its magnetic field, showed that it has at most a very small core—comprising less than a twentieth of its mass, compared to Earth's core, which makes up about a third of the planet's mass. In all, over 2,000 samples of rock, weighing nearly 400 kilograms, were brought back to Earth. These have been extensively analyzed, both for their composition and age. The oldest well-dated lunar rocks are about 4.36 billion years old—these are thought to be the solidified remains of an originally molten surface—and so the moon itself must be even older. The biggest discovery, though, is that the composition of lunar rocks is almost identical in many respects to equivalent igneous rocks on Earth. Putting all the observations together, it is clear that the moon is very similar in composition to Earth's mantle—too similar, in fact, unless the moon was originally part of Earth. The fact that the moon is slowly moving away from Earth only strengthens this conclusion. But how and when did they split up?

The most viable theory is that early Earth was involved in a planetary car crash with another, much smaller planet. The so-called impactor has even been given a name: Theia, the Ancient Greek deity who was mother of Selene, the moon goddess. Theia had to have been about the size of Mars to generate enough force in the collision to spall off the moon. Like Earth, it must have formed through clumping and accretion of the solar nebula, but with an orbit that only occasionally crossed Earth's. And there are con-straints on what happened in the collision because the moon has virtually no core. The simplest explanation is that the collision with Theia finalized the separation of the Earth's core, and most of it remained behind when the moon split off. However, it is not clear if *all* of Theia was absorbed into Earth and the moon, or whether some of it managed to escape into space. In any case, the moon most likely has an age that is very similar to that of Earth's core, within approximately 100 million years of the creation of the solar system. It has been calculated that, once formed, the moon's surface may have taken a further 100 million years or so to finally solidify. This may have included a period of remelting, caused by the energy generated from the gravitational pull of a much closer Earth—consistent with the oldest ages of the lunar crust, around 4.36 billion years.

The impact of Theia set in motion the rotation of the moon around Earth. Today the moon orbits Earth once a lunar month, and their mutual gravitational attraction gives us the twice-daily high and low tides. It is possible to estimate tidal ranges in the past from the pattern of fine layers of sediment stirred up by the tides and deposited to form sedimentary rocks called rhythmites. Rhythmites provide an extraordinary daily record of tides at various times in the Earth's history, which can be used to determine both the length of the terrestrial day and the lunar month. From this, astronomers can estimate the moon's distance from Earth at various times in the past by making use of an important physical quantity of rotating bodies called angular momentum, which depends not only on how rapidly the bodies are rotating, but also how far apart they are. The laws of physics predict that the angular momentum of the earth–moon system will have been virtually constant over time. The moon is sufficiently small that one can neglect the effect of its own spin. This means that if the lengths of the terrestrial day and lunar month have changed in the past, then the distance to the moon must also have changed, in such a way as to preserve angular momentum. And the tidal range is strongly affected by the earth–moon distance: the closer the moon, the greater the tides.

These calculations tell us that going all the way back to about 2.5 billion (2,500 million) years ago, the distance to the moon was never less than 85 percent of its current value. The effect of the moon on the tides was greater in the past, but only slightly, so that even in the earliest Archean oceans the tides were no more than half as high again as today. In other words, contrary to many popular accounts of early Earth, which suggest that the moon was much closer and generated massive tidal waves, the reality is—apart from soon after its creation—that not a lot has changed. This is consistent with the 3.2 billion year old tidal sandstones of the Moodies Group, in the Barberton Greenstone Belt, which look very much like those today. And this tidal range played a key role in the evolution of life, by creating that in-between world in the intertidal zone that life could exploit as a stepping stone out of the sea.

The effect of the moon on Earth is more profound than this, exerted through celestial mechanics. The monthly lunar orbit stabilizes Earth's spin axis, a bit like a gyroscope. Without this stabilization, the spin axis would wobble violently, periodically tipping over onto its side—leaving one half of Earth in permanent daylight and the other in permanent night. This would certainly have profoundly affected the evolution of life on Earth, because some parts of the planet would now be too hot, and others too cold, to sustain liquid water. So, it seems that the consequences of that fateful collision with Theia, right at the birth of Earth, run deep in our planet's subsequent history.

ATMOSPHERE AND OCEANS

Surprising as it may seem, it is also possible to establish a timescale for when Earth's atmosphere became a permanent feature of the planet—no longer continually stripped away either by chance collisions with asteroids or by the escape of hot, energetic gases into space. Geochemists have employed essentially the same tricks as those used to estimate the age of the core, but with a different parent–daughter radioactive pair. Iodine-129 is another one of those radioactive isotopes with a short half-life that is not found naturally on Earth, because it has all decayed away. Its former earthly existence is only indicated by the presence of its decay product xenon-129, which is a gas. And the amount of xenon-129 gas in the atmosphere today is a measure of when the atmosphere became stable. The calculation is complicated by the fact that proportions of the different isotopes of xenon (there are nine of them, of which seven are nonradioactive) have changed through time as lighter xenon is preferentially lost to space. Corrections for this indicate that the atmosphere must date back, again much like the core and the moon, to within roughly 100 million years of the creation of the solar system, although not necessarily with the same composition as today.

Putting all these timings together, we can finally say when Earth got its concentric layers. This occurred very early on, within the first few hundredths of its history. Our planet had become like the giant peach I used as an analogy earlier on, with the core at its center, encased in mantle, and then some early form of crust, and finally the atmosphere. It had, in fact, evolved toward a gravitationally stable configuration, because, in the same way that light objects float, the layers are progressively less dense as we move upward from the center. We now need to account for the final important layer: a covering of water, forming the oceans that today cover about two-thirds of Earth's surface. This water must have been a significant component of the solar nebula—hence the abundance of ice and water on many other bodies in the solar system—and some of this would have been incorporated into Earth during its accretion. When Earth was still so hot that the surface was an ocean of magma, it is likely that much of this water had already escaped into the atmosphere as vapor. Once the surface cooled and solidified, and the atmosphere had stabilized, the water could rain back down to form the first oceans. Some scientists have argued that this was not the end of the story, theorizing that more water—and possibly even life—was delivered by later collisions with icy comets or other planetary bodies.

There is a fly in the ointment when it comes to the timetable I have just outlined. The ages of the lunar rocks collected on the Apollo missions indicate volcanic activity caused by intense meteorite bombardment near the Apollo landing sites between about 4.1 and 3.5 billion years ago—this is referred to as the Late Heavy Bombardment. In the original interpretation of these ages, the bombardment occurred mainly as a devastating discrete episode—or "spike"—around 3.9 billion years ago, which, if experienced by Earth, would be expected to blow away, or significantly damage, the atmosphere and oceans. However, it may be that the lunar samples are not representative of what happened on Earth. And a reinterpretation of their ages and significance suggests that impacts were far less frequent, occurring over a protracted period of time, rather than concentrated in a single devastating spike of the Late Heavy Bombardment. The planet has certainly sustained many meteorite strikes throughout its early history: some of the spherules in cherts that formed about 3.3 billion years ago, and now found in the Onverwacht and Fig Tree Groups of the Barberton Greenstone Belt, look just like the melt droplets that would be produced by large impacts. But these impacts were clearly not large or frequent enough to blow away the atmosphere and oceans, nor melt large portions of the surface. Fortunately, we also have a record in much older rocks that show us what the conditions were like on Earth, as I will now describe.

The March of Geological Time

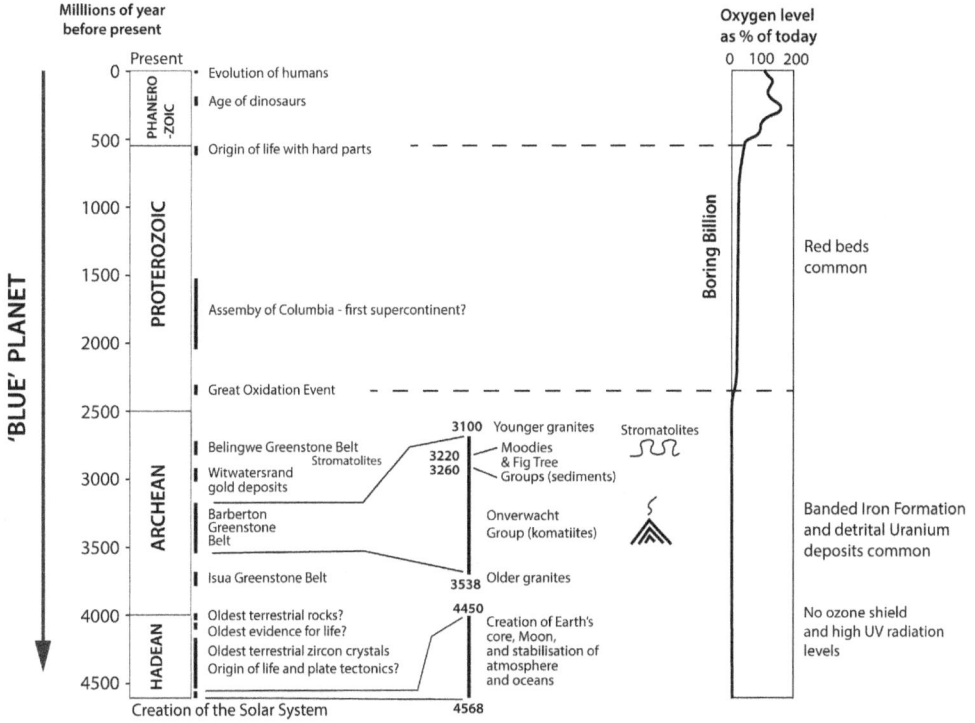

Millions of year
before present

Oxygen level
as % of today

| | | 0 100 200 |

Present — Evolution of humans

PHANERO-ZOIC

Age of dinosaurs

500 — Origin of life with hard parts

PROTEROZOIC

1000

Boring Billion

Red beds
common

1500

Assemby of Columbia - first supercontinent?

2000

Great Oxidation Event

2500

Belingwe Greenstone Belt   3100 Younger granites   Stromatolites
Stromatolites   3220   Moodies
Witwatersrand   3260   & Fig Tree
gold deposits   Groups (sediments)

ARCHEAN

3000

Banded Iron Formation
and detrital Uranium
deposits common

Barberton
Greenstone
Belt   Onverwacht
Group (komatiites)

3500

Isua Greenstone Belt   3538 Older granites

4450

Oldest terrestrial rocks?   Creation of Earth's   No ozone shield
Oldest evidence for life?   core, Moon,   and high UV radiation
Oldest terrestrial zircon crystals   and stabilisation of   levels
Origin of life and plate tectonics?   atmosphere
and oceans

HADEAN

4000

4500

Creation of the Solar System   4568

'BLUE' PLANET

FIGURE 7.3 The March of Geological Time. Our planet is just under 4.6 billion years old, and humans have existed for only a brief moment in this vast timeline. Yet for much of Earth's history, it would have looked in many ways surprisingly familiar to us, having acquired its distinctive features soon after its birth to become a world of oceans with some land, creating the "blue planet." The rocks in the Barberton Greenstone Belt provide a snapshot of surface conditions about a quarter of the way through Earth's history. Since then, the level of oxygen in the atmosphere has increased dramatically. (Illustration credit: Simon Lamb)

PUSHING THE BOUNDARIES

One of Stephen Moorbath's major contributions to geology was his work dating and understanding the rocks exposed near Isua on the west coast of Greenland. Here, in the ice-smoothed landscape at the edge of the Greenland Ice Sheet, sedimentary and volcanic rocks were found as small slivers among the granite and gneiss. Stephen published a series of papers in the early 1970s demonstrating that these rocks must be 3.7 to 3.8 billion years old, determined from the radioactive decay of uranium to lead, and rubidium to strontium. This proved that the Isua rocks are among the oldest known fragments of Earth's surface—over 200 million years older than those in the Barberton Greenstone Belt. However, unlike the latter, they were subsequently deeply buried and subjected to intense heat and pressure. It took deep erosion of the Greenland continent to expose them at the surface again.

The rocks at Isua contain clear indications about the nature of Earth's surface at this time. There are lavas with the distinctive pillow shapes of underwater eruptions, as well as possible sheeted dikes, just like the volcanic rocks found in modern oceanic crust. There are also a variety of sedimentary rocks. Sequences of banded iron formation most likely accumulated on the deep seafloor. However, conglomerates with rounded pebbles of quartz suggest either a turbulent seashore or rivers capable of eroding the bedrock. And there are hints of life: the precipitation of the iron in the banded iron formation and the buildup of faint stromatolite-like structures in sandstones may have been caused by the growth of microbial blooms, as I explained in chapters 4 and 6 when discussing similar rocks in the Barberton Greenstone Belt and elsewhere in southern Africa. The richness of this geological record has long suggested that even earlier remnants of Earth's surface exist. And in the last few decades, geologists working in Canada have recognized what might be much older rocks than those at Isua. Along the coast of Labrador, there are tiny, isolated outcrops of volcanic and sedimentary rocks—similar to those seen at Isua, but possibly at least 150 million years older (making them greater than 3.95 billion years old). These have been interpreted to be fragments of oceanic crust that were caught up in a subduction zone, or the remains of a volcanic arc above a subduction zone. Similar rocks in Canada's Northern Quebec may be even older, with a proposed age as much as 4.16 billion years old, making them relics of what geologists call the Hadean Eon (between the formation of Earth and four billion years ago). As always, whenever a group of geologists lay claim to the discovery of the oldest known rocks, the evidence for their great age is challenged. And so, it remains unclear whether these rocks really are much older than those at Isua. In any case, there is no obvious indication of the trauma of a planet-wide heavy bombardment of meteorites.

There is some widely recognized evidence for Hadean rocks, too. Zircon crystals that are about 4.03 billion years old are preserved in granites and gneiss in Canada's Northwest Territories near the Acasta River. Known as the Acasta Gneiss, these rocks may have first crystallized during the formation of an oceanic plateau derived from the melting of even older rocks, much like Iceland today. If you want to go even further back in time with confidence, then you are mainly confined to grains of sand in much younger sandstones or conglomerates. The grains are crystals of zircon—known as detrital zircons—eroded from remnants of the Earth's surface that are no longer visible or that have yet to be recognized. A treasure trove of these detrital zircon grains has been found in the roughly 3-billion-year-old sedimentary rocks at Jack Hills in Australia, which has made it possible to push back the boundaries of our knowledge of conditions on the surface of very early Earth.

Some grains contain magnetic minerals trapped inside that preserve a record of Earth's magnetic field around 4.2 billion years ago. The strength of their magnetism indicates a field comparable in intensity to that of today. This discovery shows that the magnetic field at that time may have been strong enough to protect the atmosphere from being stripped away by the solar wind, thereby making Earth a more habitable place for life. Tellingly, one grain that crystallized 4.1 billion years ago contained tiny traces of carbon with the fingerprint of life. The evidence here lies in the atomic nature of the carbon: carbon has two stable isotopes—carbon-12 and the heavier carbon-13. Living organisms can change the proportions of these two isotopes as part of their metabolism. Generally, they prefer carbon-12 at the expense of carbon-13, making the carbon in their bodies relatively "light" in terms of isotopic mass. It was this isotopically light carbon that was detected in the 4.1-billion-year-old zircon crystal, and this may be all that is left of organisms that once thrived and somehow got incorporated when the zircon first crystallized. There are indications that the same isotopically light carbon also exists in the Isua rocks. And with life there is water, indicating liquid water on Earth's surface.

The evidence is mounting that life began soon after the atmosphere had stabilized and oceans of liquid water covered much of the Earth's surface, in the early part of the Hadean Eon. I think that what is today the third distinctive feature of the Earth—plate tectonics—had already emerged too, setting our planet apart from all others in the solar system. This means that when the rocks in the Barberton Greenstone Belt were created, in the subsequent Archean Eon—spanning 4 to 2.5 billion years ago—oceans, life, and plate tectonics were already well established on Earth. I should warn you that this is a controversial idea among geologists. Even more controversially, I will argue in the rest of this book that the survival of this trinity of features depended on their coexistence, and the loss of any one of them would have eventually led to the demise of the others. The coexisting trinity also determined the course of our planet's history. So, what is the case for the very early onset of plate tectonics? Let me begin by reminding you what plate tectonics is, and how it is connected to the workings of Earth's interior.

- § -

A NATURAL HEAT ENGINE

We are used to seeing our world carved up along political lines into countries and regions of national interest. Geologists divide up the bedrock of our world in another way, recognizing a mosaic of vast plates that are constantly moving relative to each other. It is as though the Earth's surface forms an obsessively manipulated sliding puzzle—except that the pieces have irregular shapes, extend for thousands of kilometers in all directions, and have much greater freedom of movement such that they can also slide under or away from each other. And of course, by human standards, the pieces move imperceptibly slowly, similar to the rate at which your fingernails grow. This is plate tectonics, and it is mainly a description of the ocean floor that occupies about two-thirds of Earth's surface. New oceanic plate is created at spreading centers where great slabs of seafloor move away from each other, eventually plunging back into the underlying mantle in subduction zones. This way, over hundreds of millions of years, plates can move thousands of kilometers across the surface of the Earth. They are pushed at spreading centers and pulled in subduction zones, carrying the continents along with them.

Plate tectonics is a manifestation in Earth's mantle of a more general phenomenon of fluid flow called convection. As I described in chapter 3, this flow occurs in heated fluids, such as a saucepan of water on a hot stove, when the Rayleigh number of the fluid (a measure of the driving force of convection which depends on the stickiness of the fluid and how much it expands or contracts as it changes in temperature) exceeds a critical value of about a thousand. Remember that the rocks in the mantle are at sufficiently high temperatures that they soften up and become capable of flowing like a fluid, although they are still solid and incomparably more sticky than water. The Rayleigh number for the mantle is well over the minimum value required for convection, and so the fluid-like mantle rocks turn over on themselves in a pattern of convection: hot, lighter material rises from the bottom of the mantle, while colder and denser material, nearer the top, sinks. This stirs up the rocks, helping the heat in the deep interior of the mantle to escape to the surface. Mantle plumes are an important form of convection, where the rising rock is channeled into narrow tubes or stalks, with a broader return flow at the edges. However, if the rocks nearer the surface are sufficiently cold and stiff, then they can also behave as plates. The plates will eventually sink back into the mantle at a subduction zone if they are able to bend and follow the cyclic flow, and if they do not jam up—I will discuss further on what might allow this to happen. In effect, the plates hitch a roller coaster ride on the underlying convecting mantle. Today, Earth has both plate tectonics and plumes, which are really just different manifestations of the same planetary heat engine.

There seems to be wide agreement among geologists that there has been convection deep inside the Earth's mantle since very early on—mantle plumes are generally assumed to be a prominent feature of this. There is less agreement about when the convection manifested itself as plate tectonics at the surface, except that most estimates for the onset of plate tectonics fall in the first half of Earth's history. This leaves a vast period of over two billion years in which plate tectonics may or may not have been operating. Some geologists—and I include myself—side with a "business as usual," or uniformitarian, approach and think that plate tectonics began within the first half-billion years of Earth's history—in the Hadean—or at the latest, before the formation of the Barberton Greenstone Belt, given all the evidence in these rocks for plate tectonics described in previous chapters. Other geologists, however, are not persuaded by this evidence and think it started later, proposing a wide range of possible dates—up to a billion years or more after the Barberton rocks—during the middle or latter part of the Archean or even into the succeeding Proterozoic. Some have even argued that it started and stopped several times. Curiously enough, the uniformitarian view of plate tectonics is now considered the more radical one, taken as a sort of "guilty" verdict that demands a strong case.

The advent of powerful computers has opened up the possibility of simulating early Earth in a virtual world, in the hope that this will reveal how plate tectonics begin on a rocky planet like Earth. These simulations, called models in the jargon of geodynamicists, use the laws of physics to run computer calculations of how rocks flow in Earth's interior, much like weather forecasts are based on atmospheric models running inside a computer. All this requires many simplifications and assumptions about the real world. But the tricky bit with this approach is that rather like a weather forecast, which needs to be continuously updated, each model develops in a way that is very sensitive to the precise conditions in the earliest days of Earth, such as the initial temperature at Earth's surface and deep inside, and the detailed nature of the rocks in the mantle. And, of course, there is considerable debate about what these initial conditions might be!

The overall driver of what happens in these models is the way the planet eventually cools down after its hot birth, losing its heat to outer space. Starting with an early magma ocean, convection occurs in the deep interior. However, in some models, contrary to many popular notions about a hotter and more squishy outer part of early Earth, the cooling causes a strong and fixed outer layer or lid to form, a bit like the layer of brittle caramelized sugar on top of soft cream in a crème brûlée. The lid remains separate from what is happening underneath in the weaker and flowing interior of the mantle, and under certain conditions, the interior of Earth becomes hotter before beginning to cool down. It may take billions of years for the lid to break and sink into the interior, initiating some form of plate tectonics. But there are models where the lid breaks easily and plate tectonics starts very quickly.

Despite all the computing power, it is difficult to know what these models actually mean for our real Earth, because many important geological phenomena are not included in order to reduce the number of calculations involved. Models that do include more of these phenomena, such as the melting in rocks that gives rise to volcanic activity, or the way rocks change at different temperatures and pressures, or what happens when water is around, reveal widely different behaviors, and very early plate tectonics is certainly a possible outcome. As the Yale geophysicist Jun Korenaga remarked in a careful review of these models, it would be hard to use them to rule *out* the possibility of plate tectonics in the Hadean, prior to 4 billion years ago. In any case, geologists tend to be skeptical of computer models, preferring to be guided by the rocks themselves.

THE USUAL SUSPECTS

There are many things that can happen to rocks at the edges of the tectonic plates. The rocks may melt, causing volcanism, or undergo metamorphism as they are subjected to higher temperatures and pressures deeper in Earth's interior. Eventually, if these rocks remain or end up at the surface, they create distinctive belts of rocks which are well documented for much younger parts of Earth's history. The apparent absence of any of these in early Earth, or the presence of something different, is often used as "negative" evidence against plate tectonics at this time—despite the fact that other features of the rocks seem to point to a much earlier start, as I have explored at length throughout this book. Here are the usual suspects in this type of negative evidence, with my reasons why they might not be evidence after all. As you will see, most of these relate to the process of subduction.

A blue schist is a metamorphic rock that contains the beautiful pale blue crystals lawsonite and glaucophane. These crystals are known to crystallize when basalt experiences high pressures but relatively low temperatures deep within a modern subduction zone, but blue schist is not found among

Archean rocks. This has been taken by some geologists as evidence that subduction only started after the Archean, in the Proterozoic. However, in chapter 3, when discussing the origin of komatiites, I explained why the mantle must have been hotter during Earth's early existence. These higher temperatures would affect what happens to the rocks at the edges of the tectonic plates, and certain characteristic mineral transformations that take place today may not occur at all, or only very rarely. The absence of blue schist could be simply because the rocks in a subduction zone were not cold enough back then. Also, blue schist is not common even in the younger geological record, suggesting that these rocks are generally not exhumed at the surface even if they do form.

Eclogite is a very heavy metamorphic rock that forms when basalt is subjected to both the higher temperatures and pressures deep in the crust, at depths greater than about 45 kilometers. Eclogite is widely thought to play an important role in driving plate tectonics by helping the plates sink in a subduction zone. But, until recently, there were no known Archean rocks that can be classified as eclogite. It may be that almost all eclogite of this age is long gone because it is now deep in the mantle, and therefore only very rarely exposed at the surface. In fact, tiny fragments of eclogite have been found trapped inside ancient diamonds that formed nearly three billion years ago at depths of over 150 kilometers, and eclogite-like rocks of Archean age have been recently discovered in China—highlighting the fickleness of arguments based on the absence of a particular rock type. This applies in spades to my next suspect.

A chaotic mixture of rock, called a mélange (the French word for "mixture") is the product of gigantic submarine landslides on the steep and unstable edges of oceanic trenches in subduction zones. Deep inside a subduction zone, the sedimentary layers may be further broken and mixed up, caught between the tectonic plates. The apparent lack of mélange in Earth's oldest rocks has often been taken as evidence that subduction was not occurring at this time. However, closer examination of greenstone belts has led to the discovery of Archean mélange—for example, in greenstone belts in China, and, more recently, in the Barberton Greenstone Belt. In chapter 4 I described the remains of huge submarine landslides and debris avalanches preserved in the Fig Tree Group. These are, in fact, a common type of mélange and are remarkably similar to those forming today along the edge of the oceanic trench in New Zealand, set off by large earthquakes where the Pacific plate is sliding beneath the Australian plate in a subduction zone.

The geological "holy grail" in this hunt for evidence of plate tectonics in Earth's history is an unambiguous chemical fingerprint in rocks. Geochemists have proposed numerous such "fingerprints"—each with its own unique, and sometimes strange, chemical recipe—in rocks of widely varying ages, in the hope of tracking the presence or absence of plate tectonics back through time. Unfortunately, the results are almost as varied as the proposed fingerprints themselves. For Robert Stern at the University of Texas, the true signature of subduction was the abundance, over the last one billion years, of rocks derived from magmas rich in carbon dioxide (so-called kimberlites), which sometimes contain diamonds. However, Tim Johnson and colleagues at Monash University argued that marked changes in the relative proportions of sodium and potassium oxides, strontium and yttrium, and lanthanum and ytterbium, in Archean granite-like rocks, were the smoking gun for the onset of subduction around 3 billion years ago—over 200 million years *after* the youngest rocks in the Barberton Greenstone Belt had been laid down. But Blair Schoene and colleagues at Princeton University found no difference in the relative proportions of iron and magnesium oxides in basalts since 4 billion years ago, which, to them, indicated that subduction had been ongoing since the Hadean. Perhaps these inconclusive results only reflect gaps in our understanding about the geological fingerprints themselves, as well as the vagaries of finding suitable rocks to analyze. It has even been suggested that they show that plate tectonics started in different parts of the world at different times, an idea I find very hard to reconcile with our current understanding of plate tectonics as a global system.

Recently, another marker for plate tectonics was proposed by John Tarduno and his colleagues at the University of Rochester. The team built on the discovery that the magnetism of microscopic magnetic minerals trapped inside zircon crystals preserves a record of Earth's magnetic field at the time the zircon crystallized. In this way, as already mentioned in this chapter, the team found the oldest known evidence for this field. They decided to extend their study by analyzing many more ancient zircons, exploiting the fact that the strength of the field decreases overall from the magnetic poles to the equator. In this way, they thought they had a test for ancient plate tectonics: the movement of a plate over time toward or away from the equator should show up in the magnetism preserved in zircon, appearing as a decrease or increase in magnetic strength in progressively younger crystals. In fact, they found that their measurements hardly varied in zircons older than about 3.4 billion years. They interpreted this to mean that there were no moving tectonic plates before this time—in other words, that plate tectonics was not operating on early Earth.

I find many weaknesses in this evidence. To begin with, Tarduno and colleagues, despite their best efforts, could only determine the strength of magnetism very approximately, limiting their ability to determine the original positions of the rocks to within a few thousand kilometers at best. Some geologists would even argue that the intensity they measured did not relate to the ancient magnetic field, but was acquired much later on in the history of the zircon crystals. Moreover, they had relatively few of these measurements, so there were gaps of up to 100 million years during which the position of the rocks was completely unknown. As if these challenges were not enough, the magnetic data could only reveal the northward or southward movement of a plate, whereas the plates may have also moved large distances eastward or westward. So, while acknowledging the team's considerable achievement in measuring the magnetism in such ancient crystals, I think there are still too many unknowns for their results to serve as a convincing basis for dating the start of plate tectonics.

I will end my catalog of suspects with the shape and size of an Archean greenstone belt and its surrounding granites. Recall the strange pattern greenstone belts make on a geological map, described in chapter 2, in which the greenstones look very much like the leftover pastry between circular granite cookies—I am now using the term granite loosely to describe a variety of silica-rich and coarse-grained igneous intrusions. Plate tectonics, however, tends to build long linear belts of granite and volcanic rocks following the boundaries of the plates and extending for many hundreds or thousands of kilometers. Certainly, greenstone belts like Barberton and Isua are far from linear, and they are relatively small and discontinuous, surrounded by granites which have intruded and in some places squeezed their way up through the rocks at their margins. I showed in chapter 5, however, that the rise of the surrounding granites was a very late episode in the story of the Barberton Greenstone Belt and only changed the shape of its margins; the interior is quite different, largely unaffected by these granites. In other words, what we see of the greenstone belt today is a bit like pickled fruit—a fragment of the surface of early Earth preserved in granite. And the presence of so much granite does not preclude plate tectonics. Interestingly, most of the proposed later dates for the start of plate tectonics are still older than some of the younger greenstone belts, which also display the same shapes defined by the surrounding granites. Almost identical features can also be seen at the edges of much younger granite diapirs, such as one in China that formed about 130 million years ago during the age of the dinosaurs.

Clearly, the peculiar shapes of greenstone belts cannot be used as evidence against a very early start to plate tectonics. In fact, the shapes may simply be a consequence of a hotter mantle in early Earth, leading to more prolific melting of oceanic crust in subduction zones and the production of large quantities of granite, which subsequently rose as diapirs, again as I explained in chapter 5. And given that only part of the original extent of the Barberton Greenstone Belt is preserved, it is not particularly small—especially in comparison with a much younger mountain range such as the Pyrenees. I also think that there is a simple reason for the existence of small greenstone belts in the context of plate tectonics, which I will explain when I come to consider the size of the early continents.

THE ROCKS HAVE IT

Nothing in my catalogue of "usual suspects" has convincingly ruled out the existence of plate tectonics in the Hadean Eon, very early on in Earth's history. Instead, the problem is working out what it would look like on the young planet when the interior was significantly hotter, and how it would be preserved in the rocks. Throughout this book I have highlighted the similarities between the bedrock of geologically young parts of the world at the edges of the tectonic plates—particularly in New Zealand—and the Archean bedrock of the Barberton Greenstone Belt. A recurring theme in my interpretation of the ancient rocks has been the process of subduction. I will summarize the evidence for this, but first you should recall that these rocks have traditionally been divided up into three distinct groups: the oldest rocks are the "unexpected" volcanic rocks referred to as the Onverwacht Group (see chapter 3), overlain by thick piles of sedimentary rocks that make up the Fig Tree and Moodies Groups (see chapters 4 and 5).

- Subduction is one explanation for the presence of those strange mantle-like lavas in the Onverwacht Group, known as komatiites, as it offers a mechanism for their creation: the introduction of sea water deep into the mantle, where it triggers the prolific melting required to produce them.
- There are telltale signs of this water in the komatiites themselves, because salty seawater has been found trapped in bubbles inside olivine crystals, and other chemical components of the komatiites also show the geochemical fingerprint of a subduction zone.
- The existence of seafloor spreading at this time, indicated by some of the volcanic rocks, and also older volcanic rocks in the Isua Greenstone Belt, implies a local increase in Earth's surface area where the tectonic plates move apart. Therefore, some form of subduction of the seafloor into Earth's interior is needed to counterbalance this, thereby ensuring that the overall size of the planet stays constant.
- The sedimentary rocks in the Fig Tree Group are very similar to those that form on the steep edges of oceanic trenches in subduction zones.
- There is evidence for huge submarine landslides and debris avalanches that disrupted and jumbled up the rocks, likely triggered by earthquakes, just like those found in much younger subduction zones.
- The volcanic rocks of the Fig Tree Group have many of the same characteristics as volcanic arcs that overlie subduction zones today.
- These rocks have been squeezed by tectonic forces, causing the layers to become contorted into folds and offset by faults—just as occurs in subduction zones, or when plates collide and push up mountains.
- And finally, there are granite-like intrusions that can be readily explained as the result of oceanic crust melting deep within a subduction zone.

In my view, everything I have so far described makes a good case for early plate tectonics. But I think there are other strong arguments too. Today, the ocean floor, covering about two-thirds of Earth's surface, has only been in existence for less than 200 million years. This is because subduction has removed older ocean floor by taking it back down into the mantle, thus wiping the slate clean, as it were, every few hundred million years. Only the buoyant continents remain behind during this intense recycling of Earth's surface. In the early part of Earth's history, the evidence suggests that continents made up a much smaller proportion of the planet's surface, with most of it—perhaps over 80 percent—being ocean floor.

We now need to ask what happened to this early ocean floor, given that the relics of Earth's surface from this time are far too small to account for more than a tiny fraction of it. Did it just hang around for billions of years, somehow immune to the geological activity we see in the Isua and Barberton Greenstone Belts, before being subducted much later on in Earth's history? In other words, from a geological point of view, was most of its surface dead, like that of Mars, for a large proportion of the planet's history? To me, it seems far more plausible that it was geologically alive, just like the parts preserved at Isua or in the Barberton Greenstone Belt, and that there was both creation and subduction of the oceanic crust, causing a constant renewal and recycling of the planet's surface, much like we see in modern plate tectonics. In fact, the oldest known greenstone belts appear to be fragments of oceanic crust or volcanic arcs, nearly four billion years old, that have been sliced into stacks of thin slivers, typical of the fate of parts of the ocean floor today as it is scraped up into a pile of rock while sliding into the mantle at a subduction zone.

I suggested in chapter 3 that another strong argument for plate tectonics is the timetable of geological events recorded by the rocks in the Barberton Greenstone Belt. When I began my research, it was generally assumed that the rocks in the Onverwacht and Fig Tree Groups were erupted or deposited in a very short time span—perhaps only a few tens of millions of years—although the age of the Moodies Group was less clear. The new high-precision ages have changed our ideas and revealed a history that is roughly ten times as long, spanning over 300 million years. This timeline, however, makes perfect sense in terms of plate tectonics as it is manifested today. It was Maarten de Wit, in his last major paper on the Barberton

Greenstone Belt (published in 2018), who demonstrated how similar the early Earth was to geologically recent times when considering the speed at which the landscape was on the move, shifting horizontally along fault lines or rising and falling. He referred to all these things as "tectonic processes," concluding that "robust rates of horizontal and vertical tectonic processes derived from the Makhonjwa Mountains are similar to those encountered across modern oceanic regions and subduction zones" (I have omitted some technical jargon to make the meaning clearer).

Maarten's conclusion seems to have now been confirmed by the magnetic compass provided by measurements of the *direction* of magnetism (rather than just the *strength* of this magnetism discussed in the previous section) preserved in both the Barberton Greenstone Belt and similarly ancient rocks in western Australia. Alec Brenner and colleagues at Harvard University concluded that at the time the sedimentary sequences in the Moodies Group were being laid down, early continents were moving at speeds typical of today and had already drifted many thousands of kilometers. The logical inference is that global plate tectonics was fully operating at this time, although the sizes of the tectonic plates are less clear. Today, a typical plate is a few thousand kilometers across, moving across the globe at a rate of about 50 to 100 kilometers every million years; it takes a few tens to hundreds of millions of years for oceans to close up, or new oceans to form—comparable to the timeline for the formation of the Barberton Greenstone Belt.

GETTING THE PLATES MOVING

For plate tectonics to work, there must be forces within Earth that can move the plates. Without these forces, the plates are like broken-down vehicles, perpetually stuck in the same place. Today, the movements of the plates are ultimately driven by the force of gravity: a heavy plate, as it sinks back into the mantle at a subduction zone, effectively pulls the rest of the plate down with it; and the buoyant mantle, where it wells up at a seafloor spreading center, pushes the plates out of the way. These forces move the plates in particular directions and at particular speeds, working against those forces that resist the plates' motion.

Back in the 1980s, Philip England, who was then at Harvard University, developed an ingenious way to work out how much force was available in early Earth to drive the plates. He made use of estimates of the pressure and temperature experienced by Archean rocks, going back 3.6 billion years ago, that were once deeply buried and intensely squeezed. The pressure and temperature could be worked out from the chemical composition of the particular minerals in the rock that crystallized at these conditions. And because the pressure is simply due to the colossal weight of the overlying rocks, Philip could use it to determine how deep each rock sample had once been buried. Putting this together with the temperatures for a variety of rocks that had been buried by different amounts, Philip calculated the temperature increase with depth in the outer part of early Earth. Geologists refer to this as the geotherm.

The geotherm determined the force that would have been needed to squeeze and bury the very rocks Philip used to calculate the geotherm in the first place. This is because rocks become weaker as they get hotter, and so a high geotherm (that is, the rocks get hotter rapidly with depth) would indicate that less force is needed than for a low geotherm. To Philip's surprise, the Archean geotherm turned out to be very similar to that measured today for mountain ranges such as the Andes of South America, or the Himalayas, or the European Alps, regions uplifted by the forces of modern plate tectonics. In fact, the pressures alone indicated heights for Archean mountains that are more or less the same as these much younger mountain ranges. All this pointed to the same forces in the Archean, also capable of driving the tectonic plates and pushing up high mountains where the plates collided.

You may have spotted an apparent contradiction between what I have just described and the idea that the mantle in early Earth was hotter than today. Surely, if the geotherm back then was the same as today, the temperatures deep within Earth must be the same too? However, in a planet like ours where the mantle is continually stirred up by convection, this geotherm only applies to the outermost parts; the deep interior temperature varies much less with depth. Thus, typical geotherms for the outer part of Earth today are temperature increases between 5 and 25°C for every kilometer of increasing depth. But below depths of 100 to 200 kilometers, the temperature increase is much smaller, eventually becoming only a fraction of a degree per kilometer of increasing depth. In early Earth, the geotherm for the outer part of Earth must have continued down slightly deeper (only a few tens of kilometers more) beneath the mountains to eventually reach a higher average interior temperature than today. This way, there is no contradiction.

So, if plate tectonics was well underway by the time the rocks in the Isua and Barberton Greenstone Belts were created, in the earliest period of the Archean Eon, can we pin down its birth? It must have been in the preceding Hadean Eon, and I think it is the availability of abundant water on the Earth's surface that is critical to determining more precisely when this happened. This is because the forces available to drive the plates must overcome their resistance to movement. Thus, a key factor in modern plate tectonics is the lack of strength of the rocks at the edges of the plates, making it possible for one plate to slide easily past or under another plate. In other words, the edges of the plates are well lubricated. A plate must also be weakened enough so that it can bend down in a subduction zone and plunge into the mantle. Geologists and mining engineers have discovered from extensive practical experience of trying to break or drill into rocks, whether in the laboratory, mines, or oil wells, that water is the necessary lubricant that weakens rocks.

Water does this in a number of ways. It reacts with the rock to form new 'papery' minerals that can slip and slide, rather like greasy paper. Under pressure, water helps to counterbalance the huge pressures inside Earth, keeping rock masses apart so that they can move past each other with less resistance. Finally, the presence of tiny amounts of water deep in the mantle, at the bottom of the tectonic plates, may trigger the formation of small pockets of magma, creating a 'liquid' base to the plate along which it can slide. Water plays a particularly important role in lubricating the edges of the plates in subduction zones. Here, it is trapped in the pore spaces of sediments that have been dragged down with the subducting plate and smeared over the moving surfaces.

The main source of all this water must be the oceans, and so I find it hard to escape the conclusion that once oceans existed on Earth, the conditions were right for plate tectonics. This implies that the timetable for the start of plate tectonics is similar to that for the first oceans—and life—within a few hundred million years, at most, of the birth of the Earth. The big question is whether something extra was required to actually start the plates moving, like a good shake to get an old-fashioned watch ticking, or a final heave to free a bogged-down vehicle. There are lots of possibilities. The push of upwelling mantle plumes, or the push-and-pull due to the tendency for the thick piles of erupted volcanic rock to collapse, or the pull exerted by a foundering plate as it starts to sink into the mantle, could have provided the necessary forces to get the plate tectonic mechanism going. It has even been suggested that the shock of a giant meteorite impact provided suffi-cient impetus. Whatever it was, it happened on a planet that was a world of oceans.

Plate tectonics on Earth, however, did more than just start. It kept going, unlike on our nearest planetary neighbours, Mars and Venus. Even if it had once started on those planets, we see no evidence today of interconnecting chains of mountains, rifts, deep trenches, fault lines, or long chains of volcanoes—features that are typical of the plate edges on Earth. Impor-tantly, there are no oceans of water either, so critical for lubricating the edges of the plates on Earth, although it seems that Mars and Venus may have had them early on. So, both the initiation and continuation of plate tectonics on Earth comes down to the fact that conditions on Earth turned out to be just right for permanent bodies of liquid water. I will leave the question *why* the conditions on Earth were just right, but not on Mars or Venus, for the next chapter.

A GLOBAL JIGSAW PUZZLE

One major difference between early Earth and today seems to be the pres-ence of much larger continents today. But the size of a continent is not the same as the size of a plate; for a typical size of plate, a small continent only indicates that most of the plate is oceanic. This is a more important point than one might first think, because the light and buoyant continents are the main survivors in the geological record. Ancient oceans, by contrast, can only be glimpsed at in those narrow and sinuous ophiolite belts—relics of seafloor (see chapters 1 and 3) that were caught up in mountain chains during continental collision. Significantly, the extent of these belts is determined by the size of the colliding continents. If the continents were much smaller on early Earth, even though the plates were roughly the same size as today, then the record of these collisions would appear as disconnected and short segments of ophiolites. Thus, if we interpret some of the oldest greenstone belts as ophiolites, their limited lateral extent would be a natural consequence of the collision of small continents, especially as subsequent intrusions of large granite bodies would have covered up much of their original connections. As little more than an educated guess, I would say that the earliest individual continents were probably no larger than present-day Spain or Newfoundland.

There is considerable debate about when the total area (and volume) of the continents approached that of today, covering roughly one-third of Earth's surface. It has been estimated that the average age of continental crust is roughly two billion years old. However, the same average value can arise in many ways, and if this were all we knew, then we would not be able to tell whether the continental crust was created gradually over Earth's history, or during major events. In fact, the rocks in the continents point to a more gradual production, but not necessarily at the same rate throughout geological time. Like a child growing into an adult, the crust seems to have had spurts of growth, followed by more stagnant times. As I described in chapter 5, the spurts in growth might have happened through accelerated periods of melting of mantle rocks or oceanic crust to produce granite in the so-called subduction factory. One of these periods might have been right at the end of the geological life of the Barberton Greenstone Belt—around 3.1 billion years ago—when it was invaded by large bodies of granite. Hundreds of millions of years later, when continental crust was more widespread, underlying mantle plumes may have triggered outpourings of volcanic rock that covered parts of these continents, forming some of the younger greenstone belts.

Once formed, the continental crust is at the mercy of the forces of plate tectonics, rearranged and remolded by the motion of the plates. Like jostling pieces of floating wood endlessly caught up in an eddy current, or perhaps dancers in an elaborate ballet, the continents sometimes come together, sometimes split apart. We can track their movements by following the trail of new seafloor they left behind, and by using the magnetic compass preserved in the rocks, although the story gets less certain the further back in time we probe. With this in mind, we can see the basement rocks of Africa as the amalgamation of much smaller older continents, like pieces in a jigsaw puzzle, building up the larger picture of an extensive continent. The continents are brought together, ultimately colliding, when the intervening oceanic crust is carried down into the mantle in a subduction zone. This will itself inevitably lead to more freshly created continental crust through the workings of the subduction factory.

The oldest continental crust in Africa, as far as we can tell, consists of fragments within southern Africa's Kaapvaal Craton, suggesting that the craton itself formed through the collision of much smaller continents. The Barberton Greenstone Belt, together with other greenstone belts that are less well preserved or exposed, can be interpreted as all that is left of the intervening oceanic crust, forming ophiolites along the joins between these continents—the rest was subducted back into the mantle. Over half a billion years later, around 2.6 billion years ago, the Kaapvaal Craton collided with the Zimbabwe Craton to the north, welded together along the Limpopo Belt to build the Kalahari Craton. The Kalahari Craton collided with and became joined to the Congo Craton. And so on, until the continent of Africa eventually emerged. But it was a very hesitant emergence, because sometimes the assembled continental fragments broke up again, as though not properly stuck together, before reassembling in new ways. In fact, the shape of Africa we see today was carved out by the most recent phase of continental breakup.

The creation of Africa is clearly part of a global story. If we go back about 2.1 billion years ago, the evidence points to the beginnings of what might have been the first supercontinent, called Columbia. Columbia was built up through the collision and amalgamation of many smaller continents and reached its maximum size about 1.5 billion years ago. It then split apart, but the fragments eventually collided again to form the supercontinent of Rodinia about 900 million years ago. Yet again this broke up, and yet again, around 250 million years ago, the fragments had come together as Pangea, so that the world virtually became a single landmass surrounded by ocean. Subsequently Pangea split up, leaving behind the supercontinent of Gondwana, which was made up of all the present-day southern continents plus India and Arabia. And starting about 180 million years ago, Gondwana broke up, leading to the arrangement and shapes of continents that make up our world today.

The dance of the continents is just another manifestation of convection in Earth's interior, choreographed by both plate tectonics and mantle plumes. Plate tectonics has moved the continents around, but the plumes may have been instrumental in determining where and when the breakup of a supercontinent took place. This is because whenever a plume formed underneath, it pushed up the surface into a dome reaching roughly a kilometer high and several thousand kilometers across. Under the force of gravity, the dome would have the tendency to collapse and crack, splitting and stretching the crust as the supercontinent literally started to fall apart. The telltale signs of the plumes are extensive volcanic eruptions along the newly created rifts, where magma has risen up from the underlying plume head. The remnants of lava flows that erupted during the breakup of Gondwana can be found today along the western and eastern edges of southern Africa, across the Atlantic in Brazil, and in northern India.

The plume-induced breakup of a supercontinent does not just change the shapes and positions of the continents; it may also lead to the mass extinction of life. This is because the outpourings of lava along the newly formed rifts can be large enough to cause a sort of nuclear winter, darkening the skies with vast quantities of ash and sulfur. Such large eruptions can also affect the climate in the opposite way, precipitating longer-term global warming through the release of carbon dioxide, matching or even far exceeding our own industrial activity. The mass extinction of life about 66 million years ago, when the dinosaurs were wiped out, coincided with gigantic volcanic eruptions in northern India (these are the Deccan Traps, consisting of over one million cubic kilometers of lava) during the final stages of the breakup of Gondwana. However, it is now widely thought by geologists that a mass extinction on this scale was the combined effect of these volcanic outpourings and another planetary wild card: a colossal asteroid, up to 15 kilometers (9 miles) in diameter, that crashed into what is today the Yucatan Peninsula in Mexico at precisely the same time. It has been calculated that the impact threw enough debris into the atmosphere to have even worse consequences for the climate than the dust from the volcanic eruptions. Certainly, life has felt the hammer blow of sudden and drastic changes in the atmosphere and climate due to plumes beneath the continents before—consider the mass extinctions about 250 million years ago, both on land and in the oceans, when even more extensive lavas than those preserved in the Deccan Traps were erupting in Siberia, fed by an underlying superplume head.

Yet all this was not enough to extinguish life altogether. Life also endured the vast spans of time between these freak events. In the next chapter, I will look at the history of our planet's climate and the underlying causes of both a stable climate and one subject to change. In doing so, I will explore the idea that there is a close relationship between living organisms, the atmosphere, climate, and the planet itself—and this interconnectedness is why the stream of life was never fully cut off.

- § -

TIME OUT

Not all my time in Swaziland was spent camping in the Malolotja Nature Reserve. I needed a break from my field work from time to time, and it was always a rude shock when I emerged from my weeks of isolation in the reserve and interacted with the hectic world outside. During one of my field seasons, I was introduced to a United Nations nutritionist who was advising the Swaziland Government on health policies. She had a spare room in her large house, which she rented to me as a base in Mbabane when I was in town. Her house was also a meeting point for American Peace Corps volunteers. There was sometimes a party atmosphere when I got back from the field, and we would all go out for a meal or sit through a worn-out movie with distorted sound at the local cinema.

I would usually have a big shop on my first morning back in town, stocking up with supplies for my next field trip. I remember returning from shopping once to find the distinctive blue Land Rover of the Royal Swazi

Police parked outside the house. I was told that there had been a burglary, and the thieves had gone off with all my field equipment. Stupidly, I had made it easy for them by putting all this in a blue pack, including my field notebooks, maps, aerial photographs, camera, and film rolls. In effect, my entire research over the past couple of months had been lost. I was devastated. We soon worked out that the thieves had come by an overgrown field behind the house and broken in through a window that looked out on the field. When we tried to follow their tracks, we spotted the notebooks and maps tossed on the ground not far from the house. In the end, I recovered everything except my pack and camera. At this point I left for Johannesburg on my trip to the mines of Namaqualand and Namibia, described in the previous chapter. When I returned, I found that the police had even recovered these for me. I was told by the policeman that they had very quickly discovered who had been responsible because the sudden appearance of an expensive Nikon camera had created a sensation in the local community.

I have already written about the trail motorbike I bought in Swaziland on my first visit. I often rode it around Mbabane. One evening I was visiting some friends, and I decided to take a shortcut along some backstreets. These were unsealed and very rough. As I was going down a steep hill, a pack of barking dogs—there must have been four or five of them—approached me from a side street, running straight toward the motorbike. I knew from previous experience that dogs are fascinated by wheels, and they will try to bite the front tire of a bicycle. I was also worried about getting rabies if they bit me. I accelerated, hoping that I could outperform them. But one of the dogs was so fast that it managed to run right in front of the wheel. I hit the dog head on, and came flying off the bike, landing on the stony road on my right knee. I lay there in agony as the dogs stood around barking their heads off. Finally, somebody came along and scared them away. I do wonder what happened to the one I hit. It must have been injured as well, but I suppose it slunk off.

I realized when I tried to stand up that I could hardly walk. The pain was excruciating, and blood was seeping through my ripped pants. Fortunately, I was not far from the people I was trying to visit, so I managed to hobble to their house. In the end, the only way to get at the knee was to cut off the pant leg. This revealed a deep hole in my kneecap, which I filled with antiseptic cream. The next day I was taken to the medical clinic where the wound was cleaned and bandaged up. For the following three weeks I could only limp, and I spent most of my time lying on my camp bed. It was depressing to see the field season slipping away without being able to do anything. I did finally get back into the field, and I extended my stay to make up for the lost time. But from then on, I was very reluctant to use my motorbike, and I ended up lending it to a Peace Corps friend. I also have a prominent scar on my kneecap to remind me of the accident for the rest of my life.

During one of my field seasons, I learned from my Peace Corps friends that an American aid worker would be soon leaving Swaziland and they wanted to give him a going-away party. We decided that we would all meet up at the Hlane Royal National Park in the eastern part of the country and spend the weekend there. I drove down on my own in my old oil-guzzling Land Rover. The roads were unsealed, and it was very dusty. Not far from Hlane, I noticed a man standing at the side of the road. He was wearing a suit and holding a briefcase, with a newspaper rolled up under his arm. He put his other arm up as I approached him, so I stopped. He wanted a lift to the next town. I explained that I was turning off to the Hlane game reserve. He did not seem to mind this, so I waved him into the vehicle. As I drove on, he opened his newspaper and started to read, looking for all the world like a commuter on the London Underground. A little further on there was a small group of people waiting by the road. I stopped and explained again where I was going. They all got in too. This happened yet again, and before long, the car was completely packed. I realized that they were all expecting a bus that had failed to come—it had probably broken down somewhere—and I had taken over as the bus service. However, when I reached the turn-off to Hlane, there was consternation amongst my passengers. I left a very disgruntled group on the roadside.

I found my Peace Corps friends already set up at the reserve campsite, with plenty of food and beer for our evening braai. And everybody wanted to watch the waterhole at dusk. It was an extraordinary sight, as the various species of antelope gingerly came down to drink. Then the giraffes arrived, awkwardly spreading their front legs so that they could bend their necks down to reach the water. Finally, elephants turned up. Once they had had their fill, they quickly left. Afterwards, we went back to the camp and lit a big fire. It was here that I saw for the first time somebody opening a beer bottle using his eye socket. Another person did it with their teeth. Despite the potential for serious injury, these demonstrations of South African toughness only seemed amusing in the shadows of the flickering flames, as we sat beneath the wide expanse of a starry night sky, listening to the call of wild birds and trumpeting of elephants. The next morning, however, our camp was broken up by one of these elephants, which casually strolled through our tents, oblivious to the havoc it was causing.

8 | A Partner for Life

Maintaining a habitable planet: Forging links between climate, life and the workings of Earth

I t is sometimes said that the English are obsessed with the weather. I am English myself, and it is certainly true that I have spent a lot of time talking about it. There seems to be a preconceived idea of what English weather *should* be—warm and sunny in summer, but not too hot or dry; cold and snowy in winter, but not too much snow and not too cold; and spring and fall are expected to be mild, often wet and windy, but not *too* wet and windy. This collective wisdom is a perception (filtered through the vagaries of human memory) of what we would call climate, constituting long term weather patterns and surface conditions—particularly temperature—over many years or decades. We can extend the idea of climate to the whole planet, summarizing its more regional climatic zones from the freezing poles to the warm and humid tropics.

Overall, Earth's climate is very mild in comparison to the extreme conditions on other planets in the solar system. Our planet has a surface temperature averaged over its whole surface, and over many years, of about 15°C. Clearly, living organisms can survive in this climate. And the climate must have remained within the limits of what is survivable almost since the birth of the solar system, because life is nearly as old as this. One can begin to see how remarkable this is when one thinks about our nearest planetary neighbors—Mars and Venus—because their climates appear to have radically changed: there is evidence for times in the past when they might have been capable of supporting life (see further on), yet their climates are far outside these limits today. This raises the possibility of some sort of stabilizing mechanism on Earth which has kept its climate relatively mild over almost all its history.

CHICKEN OR EGG?

What is certain is that Earth's climate, in addition to being powered by the heat from the sun, is influenced by many other features of our planet. As we shall see, the oceans, living organisms, volcanism, and plate tectonics all have profound impacts on the climate. However, these impacts work both ways. If the climate is too hot or too cold, neither life nor oceans can exist. The implications of this run deep in the workings of our planet, because, as I described in the previous chapters, water is essential to plate tectonics—acting with sediment as a lubricant at the edges of plates and preventing them from jamming up. This is especially so when sea water is taken down into a subduction zone, and here it also triggers the volcanic activity in the overlying chains of volcanoes. But for water and sediment to do all this, there must be a climate that allows permanent oceans. Thus, plate tectonics and much of the planet's volcanism are also dependent on climate.

This is a bit like the notorious chicken or egg paradox: was the first egg laid by a chicken, or did the first chicken come from an egg? In other words, what is cause and what is effect? In this chapter, I will explore these ideas using the record of Earth's atmosphere, climate, and life, going back as far in time as possible. Our knowledge of surface conditions provided by the rocks at Isua and in the Barberton Greenstone Belt (our nominal reference point 3.5 billion years ago), described in previous chapters, is a key early part of this story. The aim is to work out the underlying reasons for the long-term existence of Earth's climate, a climate that makes it a uniquely habitable planet for life in the solar system. Let's begin with the most fundamental question of all: why is there life on Earth?

AN EARTHLY WOMB

The chemistry of life is based on carbon, hydrogen, nitrogen, and oxygen (plus phosphorus and sulfur). All these elements are found in meteorites known as carbonaceous chondrites, including as basic organic molecules—such as amino acids—found in living organisms. Carbonaceous chondrites are the oldest known rocks in the solar system and are thought to be relics of the solar nebula from which the solar system was created. The presence of these organic molecules must therefore be a consequence of chemical reactions at this time. In fact, this was demonstrated in 1952 during the famous Miller-Urey experiment, conducted by Stanley Miller and Harold Urey. In it, they simulated possible early atmospheric conditions—such as those involving lightning strikes in the solar nebula or Earth's earliest atmosphere—inside a glass flask subjected to frequent electrical discharges. They found that within only 24 hours, over twenty different amino acids—many of which are building blocks of life—had been created from their basic components.

The big step forward on the road to life was the emergence of organic molecules capable of reproducing themselves and carrying a genetic code that was the blueprint for the production of other molecules. This advance must have happened very early in the history of the planet, given that the evidence points to the existence of living organisms not more than a few hundred million years after the creation of Earth. Thus, it may not have been the initiation of life that was so very unlikely, but rather its subsequent survival. However, it remains unclear *where* exactly life began. Did it start in the deep ocean around hydrothermal vents? In shallow water? On land in small ponds? Deep underground? It has even been suggested that it was brought to Earth by chance meteorite or asteroid impacts, implying an origin somewhere else in the solar system—or possibly even the galaxy.

I would say that if life is very unlikely, then the existence of a widespread environment where it could start would improve the odds. The record in the oldest fragments of Earth's surface, preserved in the Barberton Greenstone Belt and Isua, suggest that oceans with ubiquitous hydrothermal vents covered much of the planet. I think this is the most likely place for the origin of life. Here, the local environment is dominated not by the atmosphere but by Earth's underwater volcanic activity, forming a sort of earthly womb. The vent acts as the umbilical cord to this womb, bringing up the basic ingredients of the chemistry of life from deep inside the planetary interior. In this case, one could truly describe Earth as our mother! The January 2022 eruption of an underwater volcano near Tonga in the southwest Pacific—Hunga Tonga–Hunga Ha'apai (or simply Hunga)—has provided a new perspective on all this. This volcano is part of the volcanic region that includes West Mata volcano, discovered in 2009 to be erupting a rare type of magnesium-rich lava called boninite (discussed at length in chapter 3). Boninite is the closest modern example of lava resembling those strange komatiites that were such a prominent type of volcanic rock on very early Earth. Komatiitic volcanic eruptions most likely took place close to the sites where life first originated. The Hunga volcano may have revealed another important aspect of these eruptions: it was so explosive that it generated a turbulent mushroom cloud of ash that burst out of the ocean and reached up into space with the energy of a 60-megaton atomic bomb.

During the next eleven hours, over 200,000 lightning strikes flashed through the cloud, creating similar conditions to those in the Miller-Urey experiment I described at the start of this section. So, perhaps the earliest organic molecules were actually made in these eruptions and rained back down into the ocean. The eruption at Hunga volcano in 2022 left behind a volcanic stump, rising nearly to the surface, with an 850-meter-deep crater in the middle. This crater is now virtually isolated from the rest of the ocean and contains both the volcanic fallout, including any newly created organic molecules, and a chemical soup derived from numerous hydrothermal vents. Perhaps we have here an example of the early crucible of life?

Millions, and possibly billions, of eruptions like this could have taken place in the first few hundred million years of our planet's life. Volcanic layers made up of komatiitic ash, such as those commonly found in the

Barberton Greenstone Belt, are abundant evidence for explosive eruptions in Earth's earliest days. These might have offered virtually endless opportunities to manufacture some of the key components of a living cell and could perhaps have initiated the chemistry of life in its earthly womb. In other words, life was likely born out of the extreme violence on early Earth—a far cry from Charles Darwin's original idea of life beginning in a benign "warm little pond"! The next developmental stage would be getting "living" stuff—or its key components—into the shallow seas and onto land so that they could interact with the atmosphere. In fact, it has been suggested that an environment that periodically dries out might be needed to create cellular membranes. This may have only taken the buildup of a volcanic edifice to the sea surface, or the creation of land as it is pushed up in a subduction zone or above a mantle plume, or even just an explosive volcanic eruption that catapulted organic material out of the deep ocean and onto dry land.

We now need to address the question of *why* conditions on Earth were suitable for the origin and subsequent survival of life, given that all the rocky planets must have been volcanically active during their early histories.

THIRD ROCK FROM THE SUN

There was an American TV comedy series called *Third Rock from the Sun* that aired about twenty years ago. The title comes from the setup, which involves a family of extraterrestrials who land on Earth and try to live like the locals. Much as I dislike taking cues from sitcoms, I think the title sums up how geologists should be thinking if they want to understand the history of our planet. Earth is part of the solar system, orbiting the sun. There are other rocky planets orbiting closer to the sun (Mercury and Venus) or farther from it (Mars). Farther out, nearer the fringes of the solar system, the planets are either gas giants or balls of ice. And being the third big rock from the sun (excluding the moon), Earth lies in what has been dubbed the Goldilocks Zone for a planet that can support life.

Such a planet has a surface that is not too hot and not too cold, so that permanent bodies of liquid water can exist. This brings us back to the idea that this water, if it forms oceans, would also make plate tectonics possible, by acting as a lubricant at the edges of the plates. And subduction, as part of plate tectonics, would then carry the water down into the mantle, triggering prolific melting and driving volatile-rich explosive volcanic eruptions that produced vast turbulent ash clouds lit up by lightning strikes—a type of eruption that may have been critical to the initiation of life, as I discussed earlier in this chapter. So, only those rocky planets that can maintain the right surface temperature for liquid water have a chance of ending up like Earth today, with its trinity of oceans, plate tectonics, and life.

It was while making the *Thin Ice* documentary about the science of climate change in 2009 that I really began to understand the fundamental physical principles governing planetary surface temperatures. This was because I had the opportunity to talk at length with Raymond Pierrehumbert, one of the world's leading atmospheric physicists. He is now a professor of atmospheric physics at Oxford University, but at the time, he was at the University of Chicago. Raymond explained the problem from the point of view of energy balance. Here, a key insight is that the hotter a body is, the more energy it radiates. And so, a planet maintains a surface temperature that ensures the energy it absorbs from the sun is balanced by the energy it radiates back into space. The heat from Earth's interior plays a negligible role in this balance, as it is so small in comparison to the solar radiation. Raymond illustrated this balance with his hands, bringing one of them toward the other for the incoming radiation, and then moving it away again for the outgoing radiation. For planets with thick atmospheres, the balancing temperature actually corresponds to the so-called radiating level in the atmosphere—the altitude where radiation can finally escape to space. The surface, meanwhile, could be significantly hotter due to the warming effect of atmospheric greenhouse gases. We will return to the role of these gases in controlling climate later in this chapter. In fact, the required temperature for the radiating level in Earth's atmosphere—which is at a height of about 5 kilometers above the surface—is only about −18°C and 33°C colder than the average temperature at ground level.

The intensity of the solar radiation decreases with increasing distance from the sun. Therefore, all else being equal, the average temperature of a planet's surface must be less for planets that are farther away from the sun in order to satisfy the energy balance of a planet's incoming and outgoing radiation. More precisely, the absolute temperature (that is, measured from absolute zero) will vary inversely with the square root of the distance from the sun. If a planet increased its distance from the sun by a factor of four, its absolute surface temperature would be halved. Venus is about seven-tenths of Earth's distance from the sun, so if it were the same as Earth in all other respects, it should be about 50°C hotter. Likewise, Mars is one and half times Earth's distance from the sun, and so it should be about 50°C colder. These calculations give a surface temperature that is averaged over day and night. However, if the planetary day is long—for example on Mercury, where the solar day is nearly two months—then the contrast between day- and nighttime temperatures can be extreme if the atmosphere, like Mercury's, is very thin.

As I will explain, the balance between incoming and outgoing planetary radiation set the background conditions that ultimately gave Earth a surface temperature that is just right for life—whereas Venus's is similar to that of molten lead, and Mars's has stayed well below zero. But there is one more thing that a planet has had to contend with in all of this: changes in the sun itself.

A FAINT SUN

The sun is a star, and so it will behave like a star. Astrophysicists have calculated that as a newly born star, during the first billion years of Earth's history, the sun was undergoing fusion with a lighter fuel, containing a higher proportion of hydrogen to helium, and so its energy output was about 25 to 30 percent less than today. Today, if everything else stayed the same, a drop in the solar output of this magnitude would cause Earth's surface temperature to plummet well below zero everywhere, and the planet would freeze over. But how can this be reconciled with all the evidence we have now for liquid water and life at an early stage of the planet's history? This is the "faint sun paradox."

It turns out that there is a relatively straightforward solution to the paradox. One lesson we have learned about Earth from our modern way of life is that the temperature of the atmosphere is very sensitive to the presence of greenhouse gases. These gases create a thermal blanket around the planet that traps the energy radiated out by Earth's surface. Technically speaking, Earth's surface mainly radiates light energy in the infrared, which we feel as heat, and it is this type of radiation that greenhouse gases are particularly good at absorbing, even when the gases are in very low concentrations. The end result is the so-called greenhouse effect which causes global warming. As the amounts of the greenhouse gases increase, the blanket becomes thicker, and the atmosphere continues to heat up. The most important greenhouse gas in the atmosphere today is water vapor, but because there is an abundance of water on Earth, the amount of this vapor is largely determined by the temperature of the atmosphere. For this reason, water vapor is regarded by atmospheric physicists as an enhancer of climate change rather than a cause of it. The next two important greenhouse gases are carbon dioxide and methane. Without these to raise the temperature in the first place, our planet really would freeze over. Assuming that, as today, there was plenty of water (and hence water vapor) on early Earth, the faint sun paradox can be resolved if there was also a higher concentration of either carbon dioxide or methane (or both) in the atmosphere—enough to produce global warming that more or less counterbalanced the reduced energy from the sun.

It might not be as simple as this. The required greenhouse effect depends on how hot you think the oceans were on early Earth. Paul Knauth and Don Lowe have argued that the oceans 3.5 billion years ago were around 40°C warmer than they are today. They based this on indicators of temperature in the detailed composition of chert layers in the Barberton Greenstone Belt, assuming these cherts had formed by direct interaction with sea water. However, Harald Furnes, Maarten de Wit, and Cornel de Ronde have concluded that the ocean temperatures were similar to those of today, also based on rocks in the Barberton Greenstone Belt. They thought they had evidence of glaciers and other signs of cool oceans. In their view, the cherts formed near local hydrothermal or volcanic vents, where very hot water was gushing out from deeper in the bedrock, and so did not provide evidence of general ocean temperatures.

Regardless of whether the early oceans were warm or cool, very large amounts of greenhouse gases would be needed. If this warming were due to carbon dioxide alone, then its concentration would have been about a thousand times higher than today for cool oceans—and much higher still for warm oceans. However, more extreme amounts of carbon dioxide seem to be ruled out by the fact that, in these quantities, carbon dioxide reacts with rocks at the surface to form certain types of carbonate minerals—and these minerals have not been detected in sedimentary rocks of this age. Very large amounts of methane would not have helped either, because at high concentrations methane forms an organic haze that limits its greenhouse effect. So, if the early oceans were indeed much warmer, there must have been something else affecting the climate.

Many suggestions have been made about what this "something else" might be, and it could even eliminate the need for large amounts of greenhouse gases at that time to explain cool oceans. One idea is that large amounts of nitrogen in the early atmosphere played a role. Today, nitrogen is the main gas in the atmosphere, comprising 78 percent of its volume—but this is under current atmospheric pressure. However, if there had been enough nitrogen in the early atmosphere to push up this pressure to a few times what it is today, then there would have been more frequent collisions between nitrogen molecules and those of carbon dioxide or methane, producing the curious effect (which I won't attempt to explain!) of increasing the amount of light energy that can be absorbed by the greenhouse gases, even though nitrogen itself is not a greenhouse gas. This way, the greenhouse effect in the atmosphere could have been boosted enough to compensate for the faint sun. The idea was tested by analyzing minute bubbles trapped in veins of quartz from ancient rocks in the Pilbara Craton of Western Australia. These contained a sample of the atmosphere 3.0 to 3.5 billion years ago. The bubbles showed that the atmospheric pressure was much the same as today, with similar amounts of nitrogen, and so nitrogen is not the antidote to a faint sun. But there is an important corollary to this conclusion, because it also demonstrates that the atmosphere back then had much the same "feel," generating similar winds, waves, and storms, and with virtually the same freezing point for water.

Another idea relies on the fact that only some of the incoming solar energy is absorbed by Earth—the rest is reflected straight back into space as though by a mirror, and so has no effect on Earth's surface temperature. How much energy is reflected is summed up by what is called the planet's albedo, measured on a scale from 0 to 1. Highly reflective ice sheets and land masses, or cloudy skies, increase the planet's albedo, while extensive, absorbent oceans or clear skies reduce it. It has been proposed that because early Earth was a world of oceans, with a significantly smaller land area than today, it would have had a much lower albedo (assuming clear skies) compared to today. If the albedo was low enough, it could have compensated for the faint sun by simply taking advantage, as it were, of more of the Sun's available energy, removing the need for extreme amounts of greenhouse gases in the atmosphere. In fact, the authors of this idea showed that it was possible with amounts of carbon dioxide only a few times higher than today's, though it would still require much more methane. As a long-term way to regulate the planet's surface temperature, however, it would have required changes in land area to have somehow kept up with the increase in solar output through time. If correct, this was certainly a fortuitous coincidence. More likely, smaller land area together with substantially greater amounts of greenhouse gases were important factors, although this still leaves open the question of why there was so much of these greenhouse gases—something I will return to later.

A faint sun must have had a profound effect on Mars and Venus too. These planets are essentially made of the same stuff as Earth, and so they must have had plenty of water vapor in their atmospheres too at the very beginning—this vapor would have needed to eventually rain down to create oceans. In fact, geologists see features on Mars, such as dried-up lake beds, vast flood plains, dry and sinuous river channels, and deep rocky canyons, which indicate that large quantities of water once flowed across the Martian surface. However, for this water not to have frozen, Mars would have required a significantly thicker atmosphere than Earth's to cope with both a faint sun and the smaller amounts of solar radiation it receives anyway. Despite this problem, the evidence for liquid water makes Mars a potential place for early life. On the other hand, for Venus to have been habitable—even with a faint sun—it would have needed a much thinner atmosphere than it has today, and maybe even a type of atmospheric solar umbrella to reduce incoming solar radiation, such as certain types of clouds capable of reflecting sunlight back into space.

Mars and Venus failed to keep any oceans they might have had. The higher solar intensity on Venus, compared to Earth, would have ultimately led to the complete evaporation of any Venusian oceans. For Mars, two other important factors came into play: its gravity and magnetic field. If sufficiently strong, both of these are capable of acting like glue, helping a planet retain its atmosphere—a strong magnetic field can prevent the atmosphere from being completely blown away by the solar wind. Mars fails on both counts; it has relatively weak gravity because it is significantly smaller than Earth, and also has a weak magnetic field. It was unable to hold on to a thick enough atmosphere, and it became too cold to sustain permanent bodies of liquid water on its surface. I do wonder, however, whether the water on Mars was ever more than just the catastrophic outpourings of melted permafrost, triggered by the heat generated during asteroid impacts. Any life that once existed on Mars (or Venus) likely went extinct long ago, although some scientists still hold out hope for life on Mars, possibly in lakes of liquid water buried under the polar ice caps.

It seems that Earth's distance from the sun was critical for determining the planet's fate. The early chance collision with Theia must have played an important role as well, leading to the formation of both the moon and Earth's core, and leaving behind a planet that was just the right size. The core allowed Earth to retain a protective magnetic field, while the orbiting moon stabilized Earth's spin axis and created the ocean tides. But what about life? What part has it played, if any? Today, as far as we can tell, none of the other planets in the solar system hosts living organisms—apart from a future visiting astronaut. Later in this chapter, I will argue that living organisms have played a significant role in moderating Earth's temperature, allowing liquid water to remain present at the surface. There were times, however, when that temperature swung dangerously upward or downward—and in some cases, it seems that living organisms have driven these changes to extremes.

BREATHLESS

The one feature of the atmosphere conspicuous by its absence in much of the previous discussion is oxygen. Today, oxygen makes up about one-fifth of the atmosphere. In fact, whenever you look out at a lush green landscape, you are also looking at a vast natural oxygen-producing factory running on solar energy. The production line in this factory is oxygenic photosynthesis, carried out by cyanobacteria, algae and the plant world. I described in chapter 4 how oxygenic photosynthesis harnesses the energy in visible light to split carbon dioxide into carbon and oxygen. The carbon combines with hydrogen from water to make hydrocarbons that are the stuff of life, allowing organisms to grow. The oxygen, however, is released into the atmosphere as a gas. If you break the stem of a waterweed, and hold it underwater in sunlight, you can see this oxygen as a stream of bubbles emerging from the broken end.

When an organism dies and starts to decay, the whole process is reversed: the hydrocarbons now react with oxygen in the atmosphere to produce water and carbon dioxide. This reaction is the tragedy of our modern civilization, because we power our way of life with the large amounts of energy it releases when we burn fossil fuels in oxygen. Many organisms (so-called aerobes) also consume oxygen through respiration, in effect burning organic matter inside themselves to drive their life-sustaining chemical reactions. All this means that the oxygen in the atmosphere results from a balance between its production through oxygenic photosynthesis and its loss through respiration, the decay of organic matter, and any other chemical reactions that consume oxygen, such as the rusting of iron.

Over geological time, this balance has changed. As life colonized more of the planet, there seems to have been a net production of oxygen, most likely driven by widespread oxygenic photosynthesis of microbes such as cyanobacteria, combined with burial of their remains so that they did not decay through oxidation. This way, the amount of oxygen could build up in the atmosphere. The geological record shows that this buildup did not happen smoothly, but in large steps. Prior to roughly 2.4 billion years ago, the virtual absence of oxygen in the atmosphere also made it possible for iron to exist in an unoxidized form. This iron was soluble in water and could be transported across continents by rivers, or wafted through the oceans, ultimately ending up in shallow seas to build up the planet's vast reserves of this metal as banded iron formations. I described in chapter 6 my visit to one of these deposits in South Africa, at the Sishen iron ore mine. The lack of oxygen also meant that iron was unable to rust on land, so the sand grains deposited by rivers had none of the reddish-brown hematite or goethite staining so typical of much younger river-deposited sediments, which for this reason are often referred to as red beds. Hence the pale-colored sandstones of the Moodies Group in the Barberton Greenstone Belt, laid down by rivers in a world before there was atmospheric oxygen. Minerals of uranium, as well as pyrite—such as those found in the ore-bearing Archean conglomerates of South Africa's deep Rand gold mines—were also more stable in this oxygen-free environment.

In case you are wondering about the red landscape on Mars, it is also due to a rusting of iron that coats rocks or is present in the soil. But why is iron rusting on Mars, given there is virtually no atmosphere—and hence no oxygen—to speak of, and presumably no life? One idea is that the rust is a remnant of a much earlier and wetter period on Mars, raising the tantalizing possibility that life once existed here, perhaps even capable of oxygenic photosynthesis. However, we are only likely to know the true reasons when geologists finally get their hands on samples of Martian soils.

THE RISE OF OXYGEN

Soon after the end of the older Archean Eon, as it transitions into the succeeding Proterozoic (Ancient Greek for "time of former life"), Earth's atmosphere starts to show signs of oxygen. The mere fact that geologists have given names to these eons suggests there is something different about the rocks: red beds appear in the sedimentary record and banded iron formations become much less common. It is now widely thought that a more precise indicator of oxygen is the nature of sulfur found in sedimentary rocks. The sulfur is part of minerals such as gypsum (calcium sulfate) and barite (barium sulfate), occurring as layers or pods like the barite in the Fig Tree Group of the Barberton Greenstone Belt.

Sulfur has numerous isotopes, each with a different atomic mass, and geochemists have routinely measured their abundance in sulfur-bearing minerals. Usually, this abundance is in proportion to the masses of the individual isotopes. However, for sulfur samples from Archean rocks, such as those from the Fig Tree Group, the results are quite different. The sulfur isotopes here have highly variable abundances with no simple relation to mass. It can be shown that this sort of behavior will occur when sulfur is exposed to a high intensity of ultraviolet light. Molecules of sulfur dioxide gas in the atmosphere—derived from volcanic eruptions—absorb this light and become unstable, splitting and recombining in ways that result in complicated changes in the proportions of the sulfur isotopes. The sulfur ultimately becomes dissolved in water, where it can combine with other elements to form minerals in sedimentary rocks.

Currently, we are shielded from most of the solar ultraviolet radiation by the ozone layer (a form of oxygen) in Earth's upper atmosphere. But this protective shield would not have existed on early Earth when there was essentially no oxygen in the atmosphere. This link between the presence of atmospheric oxygen and the intensity of ultraviolet radiation reaching Earth's surface suggests a simple explanation for the unusual abundances of sulfur isotopes in Archean sedimentary rocks: it is a consequence of very low atmospheric oxygen levels (lower than 0.001 percent of today's), and the change to isotopic abundances of sulfur typical of those today marks a significant rise in atmospheric oxygen. Detailed studies of ancient samples of barite and gypsum from well-dated rocks have shown that this transition occurred between 2.4 and 2.3 billion years ago, during what is sometimes known as the Great Oxidation Event—although there might have been very local environments before this where there were significant amounts of oxygen (each one a sort of oxygen "oasis") and in which aerobic organisms could thrive.

One possible explanation for the rise in atmospheric oxygen is that, by this time, there were extensive regions of thick continental crust with wide, shallow seas along their margins—environments where oxygenic photosynthesis in microbes could flourish, and where the dead organisms could be buried on the seabed. Perhaps, too, most of the available iron had already been carried to the sea by rivers, placing a limit on the amount of oxygen that could be taken out of the atmosphere through the rusting (that is, oxidation) of iron. The presence of rusted iron in sediments and ancient soils indicates that atmospheric oxygen rose rapidly to between 1 and 10 percent of current levels. However, the deep ocean seems to have remained starved of oxygen until around 1.8 billion years ago, when the last of the extensive banded iron formations were laid down.

For the next one and half billion years or more, oxygen in the atmosphere stayed much the same, giving rise to what is sometimes called the Boring Billion. But near the end of the Proterozoic, about 0.85 billion (850 million) years ago something changed. The level of oxygen in the atmosphere began to rise again, and complex multicellular animals emerged—events that may be connected and that eventually led to what is known as the Cambrian explosion, when organisms with hard parts or shells proliferated about 540 million years ago. By 420 million years ago, oxygen was greater than 60 percent of current levels, thereafter rising at times much higher, with evidence for forest or bush fires that left behind layers of charcoal. The reasons for the second rise in oxygen are not well understood; it is widely thought that there are links with the evolution of life, although unravelling cause and effect remains a challenge.

The rise of oxygen in Earth's atmosphere shows how life is capable of affecting our planet in important ways, in this case through the liberation of free oxygen as part of photosynthesis. This would have also caused the demise of methane in the atmosphere, because methane readily reacts with oxygen to form carbon dioxide and water. This must be the explanation for why there could have been so much of this powerful greenhouse gas during the early stages of Earth's history—before the rise of oxygenic photosynthesis in living organisms—helping to prevent the surface of the planet from freezing over when experiencing a faint sun. The rise in oxygen also influenced the course of evolution: new organisms evolved that could harness the energy released when this gas reacted with other elements, emitting the greenhouse gas carbon dioxide into the atmosphere. As we shall see, life was developing many ways to influence the planet's climate.

- § -

CLEAN AIR

Today we can observe the interaction between life and the atmosphere in real time by monitoring the gases in the atmosphere. I had the opportunity to see this for myself when I visited the Baring Head Atmospheric Monitoring Station, perched on top of a cliff at the southern tip of North Island, New Zealand. The coastline here is buffeted by some of the cleanest air on the planet, blown up from Antarctica in southerly storms. An air inlet that looks like a giant snorkel is mounted on top of a steel tower and sucks up the air, which is then analyzed for a wide range of gases. This surprisingly simple technique of sampling the atmosphere has proved very effective. There are many similar atmospheric monitoring stations around the world, carefully positioned at what are known as clean air sites so as not to be contaminated by nearby human activity. In addition to the station at Baring Head in New Zealand, there are others at the South Pole, Tasmania, Samoa, Alaska, and Hawaii (among other locations). The monitoring reveals an ebb and flow of atmospheric gases through the seasons. We can watch the carbon dioxide levels drop during the Northern Hemisphere's spring and summer, as trees put on new leaves and there is regeneration of the plant world, and then rise again during the winter as trees shed their leaves and summer growth dies. We can also see the level of oxygen rise and fall in the opposite way, as oxygen is released or taken up by plants as part of their photosynthesis or decay.

There have now been several decades of atmospheric monitoring at the clean air sites. Worryingly, all of these stations show that the seasonal cycles occur on top of a year-on-year upward trend in atmospheric carbon dioxide—famously known as the Keeling Curve, after Charles Keeling, who pioneered this research. Since Keeling first started making his measurements in 1958, carbon dioxide has increased by 28 percent. The concentrations of carbon dioxide in air bubbles, preserved in ancient ice, show that the Keeling Curve is a continuation of a trend that began with the Industrial Revolution in the early 1800s, although the rate of increase has accelerated in recent decades. It has been long realized that we are the cause of this increase, through our large-scale burning of fossil fuels. This is confirmed by the steady decline in atmospheric oxygen, which is consumed as part of this combustion.

In just a few centuries, we have returned carbon to the atmosphere, in the form of carbon dioxide, that was taken out over hundreds of millions of years by living organisms as part of their growth. When these organisms died, the carbon in their bodies was buried deep underground in layers of sedimentary rock, eventually transformed into the oil, gas, and coal we use today.

MAKING ROCKS OUT OF AIR

The creation of these huge reserves of buried carbon is an important part of what is known as the carbon cycle. The carbon cycle is the movement of carbon (in the form of carbon dioxide, methane, and other compounds of carbon) back and forth between the atmosphere, oceans and bedrock, and it is, therefore, the Earth's underlying mechanism of climate change through its influence on the greenhouse gases in the atmosphere. Clearly, we have taken over some of its workings with our large-scale burning of fossil fuels. The cycle, however, is not really a single loop, but more like the nested cogs inside a watch, intermeshed with many aspects of our planet. Another important cog in this cycle turns out to be the natural process of rock weathering.

Many rocks are chemically unstable at Earth's surface because they are made up of silicate minerals (consisting mainly of silicon and oxygen) that have crystallized at much higher temperatures and pressures deep underground. It is therefore not surprising that the landscape can become a vast natural chemistry lab in which rock silicates undergo weathering by combining with water and carbon dioxide in the atmosphere to form a very common chemical called bicarbonate. Bicarbonate is probably more familiar to most people as baking soda, which is the sodium salt of bicarbonate. Importantly, bicarbonate contains carbon and oxygen, and so its creation has resulted in the removal of some carbon dioxide from the atmosphere. It also readily reacts in water with calcium—also derived from the weathering of silicate minerals—to make limestone (calcium carbonate). When the rivers wash the bicarbonate down to the sea it is quickly taken up with calcium by marine organisms to build their shells. When these organisms eventually die they are buried on the seafloor and ultimately become limestone. Before the emergence of organisms with shells, stromatolites may have played a similar role (although perhaps to a lesser extent) by driving precipitation of the bicarbonate, in the form of calcium carbonate, to cement their mounds. On the seafloor, some of the bicarbonate also precipitates with calcium—without the agency of life—as veins in cracks or in the spaces between grains of sand or silt, bonding them together to form a hard sedimentary rock.

Weathering of rocks, therefore, results in the transfer of carbon dioxide from the atmosphere to a bedrock of limestone, in effect, turning air into rock. Where the limestone lies on the sea floor, it may eventually be taken back down into Earth's interior, riding on a tectonic plate that sinks in a subduction zone. Finally, as the subducted limestone heats up, it will release carbon dioxide which can be returned to the atmosphere in volcanic eruptions. In addition, the uplift and erosion of mountains, pushed up where tectonic plates collide, provides more opportunities for the weathering of rocks. This way, both plate tectonics and the weathering of rocks are part of a profound planetary-scale cycling of carbon that operates over geological time, exerting significant control on levels of carbon dioxide in the atmosphere and, consequently, the climate.

MILANKOVITCH CYCLES

The carbon cycle can cause climate change and also be affected by it, but there are different speeds of response either way. In the short term, even if we take ourselves out of the equation, some parts of the carbon cycle involving living organisms have responded rapidly and enhanced ongoing climate change. This way, life has played a major role in driving oscillations of the climate, over tens to hundreds of thousands of years, between cold periods, called glacials, and warm periods, called interglacials.

These oscillations are set by small variations in the way Earth orbits around the sun, either in the tilt or orientation of its spin axis, or the shape of the orbit itself. They are known as Milankovitch cycles after the Serbian mathematician Milutin Milanković, who, in the early 1900s, was the first person to calculate them going back through geological time. On their own, they would only have a small impact on climate, mainly through slight changes in the amount of solar heat that reaches different parts of Earth's surface, modifying the duration and intensity of the seasons. However, the seasonal changes set off a train of events that has consequences for the global climate. For example, if there is a consistent pattern of more snow in the Northern Hemisphere winter, but less melting in summer because the summers are relatively cool or short, then over many centuries large ice sheets start to form and expand. This drives an overall cooling of the climate because as the ice sheets spread out farther from the polar region, they create a larger expanse of white snow that reflects back more of the solar radiation into space. Shorter winters and hotter or longer summers have the opposite effect, causing the ice sheets to shrink and the climate to warm. But all these changes in the climate also affect the carbon cycle—particularly the concentration of greenhouse gases in the atmosphere—finally taking the planet into a glacial or interglacial.

Drilling into the Greenland or Antarctic Ice Sheet has made it possible to directly observe a link between the Milankovitch cycles and greenhouse gases. Long cores of ice have been extracted from the drill holes, and these contain samples of the atmosphere over the past few hundred thousand years, trapped as tiny gas bubbles in the ice. The records in the ice cores show that atmospheric carbon dioxide and methane rise and fall over time, closely tracking the Milankovitch cycles and also a rise and fall in atmospheric temperature. The reasons for this response are complex, but ultimately relate to releasing or locking up carbon in the biosphere (that part of the planet occupied by life). For example, the oceans, seafloor, and permafrost contain large quantities of dead or living organisms, as well as dissolved carbon dioxide and frozen methane, constituting a significant proportion of the available carbon in the biosphere. All this can be exchanged with the atmosphere in the form of gases of carbon dioxide and methane when the oceans and permafrost warm or thaw. The bottom line is that any warming of the climate promotes a very rapid release of more carbon dioxide and methane into the atmosphere, reinforcing the warming, whereas cooling has the opposite effect, driving a rapid drawdown of these gases and further cooling. But it only takes a small shift in Earth's orbital movement for all this to reverse and the climate to swing the other way.

We know from the evidence in the ice cores that the response of the biosphere to the Milankovitch cycles involved swings in atmospheric carbon dioxide of about a quarter of its average amount, with oscillations in global temperature of about 2.5°C, above and below a middle line. However, the record in rocks shows that all this occurred on the back of much longer-term changes in the carbon cycle, which resulted in manyfold increases or decreases in the amount of carbon dioxide, and hence large swings in climate, over tens of millions of years and involving global temperature changes up to 10°C or more. Here, the role of the carbon cycle in the biosphere seems to have been very different from the rapid response in the Milankovitch cycles, so that if allowed to continue long enough (extending over many Milankovitch cycles), it eventually puts a break on long-term climate change, thereby acting as a long-term regulator of the planet's climate.

A NATURAL THERMOSTAT

The carbon cycle also makes use of the deeper workings of Earth. Mantle convection, in the form of both plate tectonics and mantle plumes, is the engine that is ultimately responsible for adding carbon dioxide to the atmosphere by driving volcanic activity, bringing up magma that is rich in carbon dioxide all the way from the mantle. The emissions of this gas can be directly measured during volcanic eruptions. The volcanoes, if left entirely to their own devices, would slowly pump up the atmospheric carbon dioxide over millions of years, resulting in an ever-warmer climate due to the increasing greenhouse effect. This is most likely what happened on Venus, which ended up with an extremely dense atmosphere made up almost entirely of carbon dioxide and a surface temperature close to that of molten lead. Fortunately, as I have already described, Earth's carbon cycle involves long term mechanisms—typically involving living organisms—that slowly take carbon dioxide out of the atmosphere and ultimately prevent the development of a Venusian hell on Earth. These include the burial of large amounts of carbon, which produced our reserves of oil, gas, and coal, as well as the creation of limestone through the weathering of rocks on land.

The weathering pathway has a key feature when it comes to influencing the planetary climate: the chemical reactions that cause weathering are themselves sensitive to surface temperature. The higher the temperature, the more effective these chemical reactions become at sucking carbon dioxide out of the atmosphere and initiating the formation of limestone. In other words, built into the carbon cycle is a natural, long-term thermostat that can automatically regulate the climate by dialing the greenhouse effect up or down—adjusting how much carbon dioxide is removed from the atmosphere and locked up in limestone.

The weathering thermostat today, however, intimately involves living organisms, both in promoting the weathering reactions on land and driving the precipitation of the limestone in the oceans. Earlier on in the evolution of life, before organisms with shells or hard parts existed, it is likely that there would have been significant differences in the way this thermostat operated. For example, without living organisms to take up what bicarbonate was produced, it would have relied on the direct precipitation of limestone from water as bicarbonate built up in the oceans, increasing what chemists call the ocean's alkalinity. It turns out that, along with the buildup of other dissolved molecules in the oceans—particularly silica (silicon dioxide)—that are normally taken up by living organisms, this activates an alternative chemical pathway for the carbon dioxide derived from rock weathering. The chemical reaction is called reverse weathering. Rather than locking carbon dioxide up in limestone, it releases the carbon dioxide back into the atmosphere and dials up the greenhouse effect. In addition, if early Earth was mainly a world of oceans, there may have been less opportunity for extensive rock weathering in the first place, although it is now thought this type of weathering can occur on the sea floor too.

This brings us back to that puzzle of how early Earth maintained a habitable climate with abundant liquid water, when the planet should have frozen over because the sun was much weaker: the so-called faint sun paradox. As I discussed earlier, the most likely explanation is that the atmosphere had much higher levels of greenhouse gases than today, producing an elevated greenhouse effect. But clearly this climate has remained habitable, despite the subsequent increase in the solar energy reaching Earth. In other words, we have a "strong sun paradox" looking forward in time from the birth of our planet.

Could it have been life's involvement in the carbon cycle that resolved the paradox, perhaps coupled with the growth of extensive continents where large-scale rock weathering can take place? If so, life did this by finding ways to slowly take carbon dioxide out of the atmosphere and progressively dial down Earth's greenhouse effect as solar output increased (although, confusingly, life's rapid response to the Milankovitch cycles had the opposite effect in the geological short term). The rise in atmospheric oxygen from oxygenic photosynthesis in bacteria, algae, and plants dialed it down even further, particularly because it also led to the removal of the even more powerful greenhouse gas methane, which was converted to water and carbon dioxide in the presence of this oxygen.

CLIMATIC EXTREMES

The record in the rocks indicates that the planetary thermostat is far from perfect, as there have been large oscillations in the global temperature over geological time. These may simply be due to time delays before the mitigating effects of the natural thermostatic system kicks in with sufficient strength. Occasionally, the climate has been forced to extremes of habitability. Dramatic changes in climate are likely to have played a key role in evolution, favoring hardy organisms that could survive extreme conditions, and created new opportunities for life.

There have been times when the planet nearly froze over completely, turning Earth into a sort of snowball. Evidence for this can be found on many continents, such as in northern Namibia, where there are sedimentary rocks with dropstones deposited by glaciers about 630 million years ago, at a time when this region lay in the tropics. Many explanations have been put forward for this. To me, the most plausible is large-scale extraction of carbon dioxide from the atmosphere by life, occurring when most of the continents were positioned in the tropics. This ultimately led to a major drop in Earth's greenhouse effect, and the growth of large ice sheets at low latitudes. Eventually, volcanic activity returned enough carbon dioxide to the atmosphere to build up its greenhouse effect once more and lift the climate out of its deep freeze.

Climatic extremes can go the other way too. About 56 million years ago, at what is known as the Paleocene-Eocene Thermal Maximum, global temperatures were around 7°C warmer on average than today. This was during what was already an usually warm period on Earth, due to the large amount of carbon dioxide in the atmosphere—three to four times what it is today—brought about by continued volcanic activity. But it has also been proposed that there was a catastrophic eruption of methane from the seabed, derived from decaying dead marine organisms, causing a brief spike of extreme warming. The methane is usually locked up in a layer of frozen gas hydrates, but a warming ocean may have caused this to gasify, suddenly adding large volumes of this powerful greenhouse gas to the atmosphere. Computer simulations of the climate suggest that the tropics became so hot, with average temperatures of nearly 60°C, that it was no longer possible for plants to survive there—creating, in effect, an equatorial desert. But elsewhere, life continued.

Within about 200 thousand years, all the methane had been oxidized to the weaker greenhouse gas carbon dioxide, returning the climate to its less extreme warm conditions. But warm conditions slowly reached a peak again, as the level of carbon dioxide continued to rise, at the Eocene Thermal Optimum about 50 million years ago. Now, the natural thermostat seems to have finally kicked in, because the accelerated weathering of the bedrock during these hot times would have sucked up large amounts of atmospheric carbon dioxide, which would have been ultimately turned into limestone by marine organisms. Thus, the amount of this greenhouse gas in the atmosphere started to come down. Then plate tectonics intervened, pushing up the mountain ranges in the Andes, the Himalayas, and Tibet over the last 50 million years. Mountain-building on this scale created new opportunities for the erosion and weathering of once-deeply buried rocks, again sucking up atmospheric carbon dioxide from the atmosphere. This drove further cooling of the climate, ultimately resulting in the Pleistocene Ice Age.

Compared to the magnitude of climate change in the geological past, the temperature rise we have experienced so far today due to human-made global warming looks small, although it is taking place extraordinarily fast. In fact, global warming is reversing the long-term climatic cooling over the past couple of million years, and so one could argue that it is an example of life playing a counter-balancing role in climate change. In any case, I don't think global warming is creating conditions that Earth has not experienced before, although it is predicted that if allowed to continue unchecked, we could end up with a climate comparable to the warmest periods of our planet's long history. I think we will manage our emissions before this very extreme situation develops. The main problem is that we, together with many other life forms, are adapted to a cooler and more stable climate, and our sheer numbers make us vulnerable to any sort of change, especially one as rapid as current global warming. There is nowhere else to go—except another planet.

I seem to have come full circle, right back to the famous *Earthrise* photograph I described at the very beginning of this book, taken by the Apollo astronaut William Anders in 1968 from the lunar orbiter. He was looking back at our vibrant and life-giving planet from the drab and sterile moon. To my mind, it is a powerful statement of a simple truth: the only truly hospitable place for us is our own home on Earth. We must learn not to make a mess of it.

THE LIVING PLANET

At the end of our long journey through geological time we have found that life is still facing the same challenges it always has when confronted with a changing world. "Same as it ever was," to quote a Talking Heads song! In the future, many species will go extinct, just as they have done in the past— this may be the fate of humans soon too, geologically speaking—but new species will inevitably emerge in new environments. This constant turnover of species is part and parcel of the stream of life, and this stream has never been cut off since life first emerged soon after the birth of Earth.

We began the previous chapter at the point when our future Earth was no more than star dust, at the birth of the solar system. Soon afterwards, in the first few hundredths of its long history, it turned into a planet we would easily be able to recognize. I have argued that plate tectonics was soon a dominant force, and along with mantle plumes has remained so ever since. In previous chapters, we have seen how these forces helped create the seascapes and landscapes of the young Earth—about a quarter of the way through its history—the record of which is now preserved in the Barberton Greenstone belt. But an essentially unchanged driving mechanism does not mean no change at all, and we see change in many aspects of Earth's history. We have witnessed on our journey the gradual emergence and rearrangement of the continents; the rise of oxygen in the atmosphere, coupled with the evolution of life; the birth and death of oceans and mountain ranges; volcanic eruptions and earthquakes; and the swings in climate between ice ages and a hothouse planet. And very occasionally, there have been brief and violent encounters with large extraterrestrial bodies.

The important thing is that Earth remained habitable and life survived. As we have seen, some of the reasons for this are intimately related to the activities of living organisms themselves—or have life's fingerprints all over them. This brings me back to that wonderful Siswati phrase for blue—"Green like the Sky"—that I introduced near the start of this book as a subtitle or epithet for Earth, perhaps engraved on a plaque on any future space mission seeking out alien life. To me, it is perfect shorthand for the many interconnections between life and the inanimate parts of the planet, interconnections that characterize our world today. So, is this just the famous Gaia hypothesis, put forward by James Lovelock and Lynn Margulis in the 1970s and named after the Greek goddess of Earth, mother of all life? Lovelock and Margulis proposed that the interactions between living organisms and Earth have created a sort of superorganism, which they called Gaia. Gaia looks after the world through its self-regulation, maintaining optimal conditions for life. In other words, once life existed, there was almost an inevitability that it would survive because its very existence keeps Earth habitable.

The Gaia hypothesis has proved to be deeply satisfying to those seeking reassurance about the state of our planet. It gives rise to the hope that in the face of human-made climate change, the rest of the natural world will step in and fix the problem. Unfortunately, the Gaia hypothesis itself has many problems. Leaving aside the issue of whether living organisms really do optimize the world for their own needs, it is difficult to see how life could have achieved this goal, turning Earth and everything on it into a single self-regulating organism. The only plausible mechanism for the evolution of life is natural selection, which works on populations of organisms by selecting those characteristics that increase an individual's chances of surviving long enough to reproduce and pass on their genes, and hence characteristics, to the next generation. But there is only one Earth—not a population of alternative Earths from which the one that best regulates its environment can be selected and go on to spawn a new population of Earths! Therefore, I think it makes more scientific sense to abandon the Gaia hypothesis as originally conceived, but use the idea of Gaia in essentially the same way as I proposed for "Green like the Sky": a form of shorthand or epithet for all the different ways that life affects the rest of the planet, whatever these effects might be. Perhaps we should call this view of Gaia by the acronym GLS, for **G**reen **L**ike the **S**ky.

We can then ask the question: would Earth have remained habitable if we took GLS out of the equation? I would say that the answer is still no. In fact, it could have been life's ability to mediate the extraction of carbon dioxide from the atmosphere, thereby reducing the planet's long-term greenhouse effect, that helped maintain habitability in the face of a strengthening sun. This does not mean, however, that GLS had any overarching control or direction. Instead, it provided a range of tools in Earth's toolkit to remain habitable (with more settings to read about in its operating manual), adding much greater flexibility to the planetary responses to environmental change and thereby improving the odds of a favorable outcome for life. These tools exist because of the extraordinary diversity of biochemical reactions in a living cell, and much of biochemistry is common to cells from different organisms. Many of these reactions are very sensitive to temperature and some involve the greenhouse gases that play a role in determining the climate. Although catering to the local needs of individual species, there were some biochemical reactions that turned out to have wider consequences. But I don't think that GLS, expressed through the chemistry of life, was enough on its own to maintain Earth's habitability; in fact, sometimes it seems to have made matters worse.

There has always been nonbiological activity going on in the background, and this appears to have provided the crucial safety net for Earth's habitability when GLS was on the wrong track. Consider, for example, how the regular cycles in Earth's orbital movement—the Milankovitch cycles—limited the extent to which living organisms could drive extreme cooling or warming of the climate during glacial and interglacial periods. Or when ever-present volcanic activity reversed a massive drawdown of carbon dioxide—most likely, in my view, due to carbon uptake by living organisms—that nearly caused the planet to freeze over. Or the converse situation, when converging tectonic plates pushed up the great ranges of the Andes, the Himalayas, and the Tibetan Plateau, accelerating the weathering of rocks and driving a continued drawdown of carbon dioxide after the hothouse conditions of the early part of the Eocene Period.

So, in many respects, life has relied on the workings of Earth, and has been shaped by them, too. But like an ensemble cast in a play, both living organisms and the geological processes of Earth have shown a sufficiently varied repertoire of behaviors to make possible life's survival over virtually the entire history of the solar system. We should, however, never underestimate luck. The fact that volcanic activity, asteroid impacts, or other changes in the environment never completely overwhelmed the world and wiped out life does not imply that the potential never existed. Whether or not life will continue to be so lucky is impossible to predict with certainty, although I would say that it is far more likely that humans will be responsible, one way or another, for our more immediate future.

CHINESE WHISPERS

Whatever the reasons, I find it an amazing fact that rocks—or the stuff they are made of—ultimately gave rise to life. This life played a role in shaping Earth's long history. But more incredibly, life, in the form of *Homo sapiens*, was eventually able to work out this history and look back over the eons of geological time, even to the extent of writing it down in books such as this so that other humans might know about it too! So, let me end my narrative with what I think is an intriguing perspective on this last thought. As we have seen in the previous chapters, geologists routinely describe events (sometimes in great detail) that occurred several billion years ago, long before humans existed on the planet. However, I am always struck by how little we know about events in our own history which took place only a few hundred years ago, and certainly in the last few thousand years.

You might think the ability of humans to communicate with each other would make it easy to acquire this relatively recent knowledge. My grandfather once told me that when he was a child, he met an old man who claimed to have heard the news when it first came out about the Duke of Wellington's victory over Napoleon Bonaparte at the Battle of Waterloo. I remember having a long and detailed conversation with my grandfather about whether this could have been possible. The Battle of Waterloo was in 1815; my grandfather was born in 1900. Assuming the old man he encountered was seven years old when he heard about Waterloo, and my grandfather was seven himself when he was told about this, then the old man must have been 99. So, it is just possible. I have often mused about this and wondered how far back in time one could go and still get reliable information about the past through a chain of oral communication.

If we take the case described to me by my grandfather, then you could say I heard something (although not much!) about the battle of Waterloo secondhand. Going back 250 years, the best we can hope for is to find out about things third- or fourthhand. Reaching back to around the time of Christ, we might hear a description of life in Jerusalem many tenths of hands removed; and by the time we reach Archbishop Ussher's date for the creation, we are over a hundredth hand away. This is a bit like the old party game known to children and adults alike in Britain as chinese whispers, but probably more familiar to those in North America as the telephone game! There seem to be far too many links in the chain of verbal communication for even the last few hundred years to be accurately remembered this way. Fortunately, we have developed other means to preserve our past. Written eyewitness accounts of people, places, and events—sometimes with pictures—are clearly the key to recording historical events, forming a long-lived repository of information that brings to life brief moments of the past.

One such example is the record of Julius Caesar's invasions of Britain in 55 and 54 BCE, over two thousand years ago. The only reason we know anything about this is because Caesar sent letters home while the invasion was still taking place and he also wrote a famous book (commonly referred to as Caesar's Gallic Wars) which I have read myself. Without this extremely unusual documentation, we would have little idea about the early history of Roman Britain. Unfortunately, because Caesar's writing is a model of terseness and his descriptions are so generalized, it is often difficult to determine what or where he is referring to. Archaeologists still argue about Caesar's landing place on the coast of southern England, or his point of departure in France. And it is sometimes unclear whether he has altered the truth, indulging in notorious alternative facts, for political purposes. Despite these limitations, his written communications are a sort of oasis of information, compared to virtually nothing before and for a long period afterwards.

We are limited to the past five and a half thousand years for any written records at all—not far off, in fact, from Archbishop Ussher's estimate of the age of Earth (!). Even so, written history more than a thousand years old is often fragmentary, with many significant gaps. Rather disappointingly, some of the earliest writing consists of little more than long lists of items. Written documents have also proved to be very vulnerable. It is sometimes said that the final destruction of the Great Library of Alexandria, in the third or fourth century CE, wiped out most of the literature of the classical world. However, the loss of many of these works may simply be due to a loss of interest in classical literature when the Roman world collapsed.

Cultural traditions preserved in dance, song, and rock paintings— such as those of the Australian Aboriginal peoples—may be more enduring and could potentially reach back tens of thousands of years, yet they can only give us no more than a faint echo of what actually occurred. Archaeologists try to fill the gaps by piecing together some of the story from their excavations of past human activity. However, most of what we leave behind is perishable, and it is estimated that only 0.1 percent to 1 percent of what survives is ever recovered. In any case, archaeological finds are usually difficult to make sense of, except in the rare circumstances where a living example or historical account gives them wider meaning. I could go on listing yet more problems in reconstructing even our more recent past. So, I can only marvel at what geology has made possible, providing us with such rich images of Earth throughout most of its long history.

- § -

MAARTEN

If you have read this book all the way through, you will have found that Maarten de Wit crops up on many occasions. Maarten was both my friend and a highly influential mentor. When I speak to Isabelle Paris and Cornel de Ronde, who also worked with him as students, it is clear that they feel the same way. Maarten is no longer with us—he died of natural causes in April 2020 during the global Covid lockdown. It was very distressing not to be able to properly celebrate his achievements in person, but the Nelson Mandela University in Port Elizabeth organized a virtual memorial, which was streamed online. This is how I managed to reconnect with Isabelle after over thirty years—her name appeared as one of the participants. Cornel is a New Zealander, but it wasn't until news of Maarten's death reached me that we decided to meet up more often to talk about our time working in the Barberton Greenstone Belt. Cornel, in fact, spent nearly eight years in South Africa, so he had lots to tell me. I decided to write this book in memory of Maarten, in the way he would have appreciated most: by focusing on the rocks and what they reveal about our planet's history.

He was a tall Dutchman with a beard. He had a habit of standing with legs apart, his hands in his pockets and leaning slightly forward, as if he wanted to catch every word you said. I think he was nearsighted; he wore glasses, but would tilt them up when he was looking closely at a rock. Dutch was his first language, so he spoke Afrikaans too. And his English was also perfect because he had been an undergraduate at Trinity College, Dublin—and at any rate, all Dutch people seem to speak perfect English! But he definitely had a recognizable Dutch accent. One of his favorite words when talking about geology was "huge." Everything was huge: a huge fault, a huge earthquake, a huge sequence of sedimentary rocks. Because Maarten tended to not fully pronounce the "h," it sounded more like "yuge." His language could be very colorful when he was excited about something. And he had sharp eyes. He would spot things in the rocks that nobody else had noticed.

So why did Maarten have such a profound impact on so many people? I think this can be summed up in one word: enthusiasm. Maarten never did anything half-heartedly, and his enthusiasm ran deep. What you were

FIGURE 8.1 A geologist in his element. Maarten de Wit, photographed in 1996, consults his geological map in the heart of the Barberton Greenstone Belt, South Africa. (Photo credit: Simon Lamb)

doing, saying, or thinking, was important to him. And this sense of importance tended to rub off, so it made you feel important too. He was generous with everything—his time, his ideas, and ways of helping you. After my terrible accident in Swaziland, when the van I was driving rolled off a precipice, Maarten made me feel that it was worth carrying on. He cared about the future of South Africa, and he was prepared to devote his energies to raising the standard of education for all.

Maarten was very well travelled—there seemed to be few places that he had not been to. And he brought all this experience to bear in making sense of rocks and what they tell us about the deep history of our planet. For him, this was more than just an intellectual exercise. It was a way to understand how our planet works, helping us all cope with the future—what he called earth stewardship science. Maarten used to say that the best geologists have seen the most rocks. In other words, if you want to understand the natural world, then you need to observe it closely.

Selected Glossary

Anticline (or Syncline) an *anticline* is a fold in layered rocks where the **Strata** or **Bedding** typically dip outward to form an upside-down "V" or "U" shape, with beds becoming younger *away* from the middle of the structure. A *syncline* is a fold in the layers where the beds typically form a "V" or "U" shape and they become younger *toward* the middle of the structure. If the original orientation ("way up") of the beds is unknown or unclear, the more general terms *synform*, and *antiform* are used.

Archean Eon of geological time between 4 and 2.5 billion years ago.

Accretionary lapilli balls of volcanic ash, about 1–10 mm in diameter, that clump together in a volcanic ash cloud and typically have internal concentric layers.

Acidic rock igneous rock with a silica content greater than 65 percent by weight; **Granite** is a typical example.

Alaskite a light-colored and typically coarse-grained igneous rock composed primarily of **Quartz** and alkali **Feldspar** with almost no dark minerals; alaskite forms a major source of the world's uranium.

Amphibole silicate mineral that tends to form needle-shaped crystals, contains iron and magnesium, and is typically green, brown, or black in color.

Andesite a fine-grained **Volcanic Rock**, named after the volcanoes of the Andes, composed mainly of plagioclase **Feldspar** and **Pyroxene**, sometimes containing **Quartz**, **Amphibole**, and **Olivine**, and classified as **Intermediate** in composition.

Anomaly departure of some measured physical property (e.g., strength of magnetic field or gravitational acceleration) from its expected value.

Asthenosphere the weak part of Earth's mantle immediately underlying the **Lithosphere**.

Banded iron formation a sedimentary rock consisting of thin (centimeter scale) alternating layers of **Chert** and iron oxides such as magnetite and hematite; they are mainly confined to rock sequences more than 1.8 billion years old, and constitute a major source of the world's iron.

Barite a rock mineral composed of barium sulfate, often occurring as pods or thin layers that formed by precipitation from water in hot springs.

Basalt a fine-grained **Volcanic Rock** composed mainly of plagioclase **Feldspar**, **Pyroxene**, and sometimes **Olivine**, and classified as **Basic** in composition.

Basic rock igneous rock with a silica content between 45 and 55 percent by weight; **Basalt** is a typical example.

Batholith a large body of coarse-grained **Igneous Rock** that intruded and cooled at depths of several kilometers or more in Earth; **Granite** typically forms batholiths.

Bedding layering in sedimentary rocks, often referred to as strata or beds, caused by variations in the nature of the sediment, defining the approximate original horizontal.

Black or white smoker column of hot, muddy water that gushes out of the seafloor.

Boninite a rare, basalt-like but magnesium-rich volcanic rock that erupts along subduction zones associated with seafloor spreading; first found in the Bonin Islands, south of Japan, hence the name. Boninites are the closest modern example of **Komatiite**, particularly komatiitic basalts, which were a type of volcanic eruption common in early Earth.

Boring billion informal term used to describe a time in Earth's history, roughly between 1.8 and 0.8 billion years ago (in the **Proterozoic Eon**), characterized by relatively stable environmental conditions and slow biological evolution.

Cambrian division of geological time between about 541 and 485 million years ago; rocks from this **Period** often contain the first evidence for fossilized life with preservable hard parts such as shells or skeletal material, once thought to mark the first signs of life on Earth, although we know now that life began billions of years earlier.

Cenozoic the most recent geological **Era** whose name means "time of young life," spanning from 66 million years ago to the present.

Continental Collision what happens when two separate regions of **Continental crust**—that is, two separate continents—come together (collide) in a **Subduction Zone** as part of **Plate Tectonics**; this usually leads to the rise of a mountain range, but it can also result in the emplacement on land of **Oceanic crust**, forming an **Ophiolite**.

Continental crust outer layer of the solid earth beneath the continents, typically 30 to 45 kilometers thick (but up to 90 kilometers beneath mountainous regions) and rich in the minerals **Quartz** and **Feldspar**, and an average composition with about 60 percent silica by weight.

Chert rock made up of a mosaic of small **Quartz** crystals (polycrystalline quartz), sometimes formed from the silica tests of **Radiolaria** in Phanerozoic rocks.

Chondrite class of meteorite made up of compressed matter from the **Solar Nebula**, internally organized into spherical bodies called *chondrules*; the oldest examples are carbonaceous chondrites, which also contain carbon and water.

Convection pattern of flow in a fluid that has a variable density, so that under gravity, less dense parts rise and denser parts sink; typically occurs in a heated fluid in which density is controlled by temperature.

Columbia a hypothetical **Supercontinent** that existed about 1.5 billion years ago, thought to be the earliest supercontinent.

Core central part of Earth, composed mainly of iron and nickel, where its magnetic field is generated; the inner core is solid, but the outer core is molten.

Craton large stable block of Earth's **Crust** forming the nucleus of a continent; from the Ancient Greek *kratos*, meaning "strength."

Cretaceous a division of time between about 143 and 66 million years ago, comprising the third and last **Period** of the **Mesozoic Era**, brought to an abrupt end by a mass extinction of life (including the dinosaurs) due to a violent meteorite impact and vast volcanic eruptions.

Cross-bedding a distinctive form of layering in sandstones, consisting of numerous layers set at angles to one another, with some truncating others, created by the internal reorganization of the sand in a migrating **Dune** or **Ripple** and laid down by either wind or water currents; it can reveal the original **Way Up** of the beds and also the direction of the currents that deposited them.

Crust outermost layer of solid Earth, differing in composition from the underlying **Mantle**; with two distinct types called **Continental Crust** and **Oceanic Crust**.

Debris Avalanche a type of landslide involving the collapse of rock and scree on a steep slope; can occur on land or underwater (submarine debris avalanche)— the latter may also result in the formation of a **Turbidite**; common on the steep underwater edges of the continents or edges of **Oceanic Trenches**, triggered by the violent shaking during an **Earthquake**.

Deformation the process or result of movements in Earth's **Crust** and **Mantle** that distort rocks by faulting or folding (see **Fault or Fold**), or changing the shape of individual crystals.

Diapir a body of rock (from the Ancient Greek for "to pierce through") that forcefully rises up through the overlying rocks, driven by its buoyancy; can be solid **Igneous Rocks** such as **Granite**, or even salt.

Diatreme an explosive and gas-rich volcanic eruption that has typically punched its way through the bedrock from more than 150 kilometers in Earth, leaving behind a pipe of volcanic fragments that may contain diamonds.

Dike an **Igneous Rock** that forms a near vertical layer due to the intrusion of **Magma** along a crack in the bedrock; typically made up of **Basalt**.

Dune mound of sand with a regular shape, typically with a steep lee face, created by the flow of water or wind.

Earthquake vibration of Earth caused by a release of energy over seconds to minutes, emitting seismic waves, and most commonly caused by sudden slip along a **Fault**

deep in the bedrock, as a result of elastic rebound when rocks reach their breaking point.

Eclogite a heavy **Metamorphic Rock**, typically rich in garnet and formed when **Basalt** is subjected to high temperatures and pressures (at depths greater than about 45 kilometers) when it is carried back into Earth's mantle by a sinking tectonic **Plate**.

Eocene a division of time marking the second **Period** in the **Cenozoic Era**, between about 56 and 34 million years ago; the climate at this time was unusually warm with high levels of carbon dioxide in the atmosphere.

Eon the largest division of geological time, typically spanning hundreds of millions to over a billion years. The current eon, the **Phanerozoic**, was preceded (from oldest to youngest) by the **Hadean**, **Archean**, and **Proterozoic** eons.

Era a subdivision of an **Eon**, typically lasting several hundred million years. The current **Eon** (**Phanerozoic**) includes three eras, from oldest to youngest: the **Paleozoic**, **Mesozoic**, and **Cenozoic**.

Faint sun paradox a problem created by the fact that during the early part of Earth's history the sun was emitting 25 to 30 percent less solar energy, which would be expected to cause the planet to freeze over, yet there is abundant evidence for liquid water and oceans at this time; the most likely solution is that the atmosphere had higher levels of greenhouse gases, enough to cause a global warming that would counteract the "fainter" sun.

Fault a break in Earth's **Crust** along which sudden movement will trigger an earthquake; can be categorized as: 1) a reverse or thrust fault where movement results in "squeezing" of the rocks, 2) a normal fault where movement results in extension of the rocks, and 3) a strike-slip fault where rocks slide horizontally past each other.

Feldspar a silicate mineral with aluminum and other elements, common in igneous rocks, particularly **Granite**; divided into the alkali and plagioclase feldspars, with many further subdivisions.

Fluid substance capable of flowing; under certain conditions, even solid materials can behave as fluids.

Fold bent, tilted, or twisted layers of rock; see **syncline** for definitions of different types of folds.

Gabbro a coarse-grained and dark intrusive **Igneous Rock** with a similar composition to **Basalt**, mainly made up of the minerals **Pyroxene** and plagioclase **Feldspar**, which has cooled slowly deep in the **Crust**.

Gaia hypothesis a controversial idea put forward by James Lovelock and Lynn Margulis in the 1970s proposing that Earth behaves as a single "organism" that regulates itself to ensure optimal conditions for life.

Garnet silicate mineral with a typical red color; forms equidimensional crystals in an **Igneous Rock** such as **Granite**, or **Metamorphic Rock** such as **Eclogite**.

Glaucophane a typically blue-colored rock mineral that forms in a **Metamorphic Rock** at high pressures, but relatively low temperatures, such as when **Basalt**

recrystallizes as it is carried down with the sinking tectonic **Plate** in the shallower
parts of a **Subduction Zone**.

Gneiss **Metamorphic Rock** derived from either **Igneous Rocks** or **Sedimentary Rocks**
at high temperatures and pressures and is made up of coarse crystals arranged
as alternating bands or thin layers of different colors, alternating between light-
colored minerals such as **Quartz** and **Feldspar**, and dark-colored minerals such as
Mica (biotite) and **Amphibole**.

Gondwana a **Supercontinent** made up of what are today the southern continents
together with India and Arabia, that separated from **Pangea** about 200 million years
ago, and started to fragment itself soon afterwards, around 180 million years ago.

Granite a coarse-grained intrusive **Igneous Rock** commonly found in the **Continental
Crust** with roughly equal proportions of **Quartz**, alkali **Feldspar**, plagioclase
Feldspar, with some **Mica**; classified in composition as **Acidic**.

Granodiorite a coarse-grained intrusive **Igneous Rock**, similar to **Granite** but with
more plagioclase than alkali **Feldspar** and more dark-colored minerals; common in
the **Cratons**.

Gravity the effect of the mutual attraction between masses, causing bodies to
accelerate toward each other; gravity on Earth is quantified in terms of the
gravitational acceleration (g).

Great Oxidation Event the point in Earth's history when the level of oxygen in the
atmosphere became significant, 2.4 to 2.3 billion years ago, rising from below 0.001
to between 1 and 10 percent of today's oxygen levels.

Greenhouse gases gases that are effective at absorbing heat; some important
greenhouse gases in Earth's atmosphere include water vapor, carbon dioxide, and
methane.

Greenhouse effect heating of air masses, such as planetary atmospheres, through the
trapping of heat by greenhouse gases.

Greenstones (greenstone belt) an old mining term for ancient sequences of
volcanic and sedimentary rocks within the **Cratons** that have experienced intense
Deformation, typically surrounded by intrusive **Granite**; the term *greenstone belt* is
more commonly used by geologists.

Hadean the **Eon** of geological time from the formation of Earth, about 4.6 billion to
4 billion years ago.

Igneous rock rock that has cooled from a molten state; it may lie within the **Crust**
(intrusive rock), such as a **Dike** or **Batholith**; or erupt at the surface (**Volcanic
Rock**), as a **Lava** or accumulations of volcanic ash made up of fragments of
solidified **Magma**.

Intermediate rock an **Igneous rock** with a moderate silica content (55–65 percent by
weight); **Andesite** is a typical example.

Isochron (dating) a technique of dating rocks that is based on **Radioactivity** and
involves fitting a straight line (the isochron, meaning "same age") to the ratios of
Isotopes from a particular radioactive decay system, measured in different samples
of the rock to be dated.

Isotope one of two or more forms of a chemical element that have the same atomic number (number of protons) but different atomic masses (number of protons and neutrons).

Isostasy the principle that Earth's **Crust** floats on a fluid-like **Mantle** in equilibrium, like an iceberg or ship at sea, according to Archimedes's principle.

Jurassic the middle **Period** of the **Mesozoic Era**, between about 201 and 144 million years ago, when dinosaurs were a prominent part of life on Earth.

Kerogen remains of organic matter (primarily dead plants and algae) in sedimentary rocks, often forming a dark residue under the microscope; when heated up at depth in the **Crust**, it is the primary source of the world's deposits of hydrocarbons such as oil and gas.

Kimberlite pipe a pipe formed by an explosive volcanic eruption called a **Diatreme**. Named after the town of Kimberley in South Africa, where the diatremes are exceptionally rich in diamonds.

Komatiite a type of ultramafic mantle-derived volcanic rock defined as having crystallized from **Magma** containing at least 18 percent by weight magnesium oxide (MgO), with low silicon, potassium, and aluminum, and high to extremely high magnesium content. Rocks that are closer to **Basalt** in composition, although still unusually rich in MgO, are called *komatiitic basalts*. True komatiites tend to form sheetlike and runny **Lava** flows, whereas pillow shapes are typical in the "stickier" komatiitic basalts.

Lava molten rock (or its solidified remains) that flows from a volcano during an eruption, forming **Volcanic Rock**.

Lawsonite a rock mineral that is sometimes pale blue and forms in **Metamorphic Rock** at high pressures, but relatively low temperatures, under similar conditions and circumstances to **Glaucophane**.

Lithophile an element that is typically found in the rocks of the **Continental Crust**, and therefore classed as a rock lover (from the Ancient Greek *lithos*, meaning "rock," and *philos*, meaning "love").

Lithosphere the strong outer part of solid Earth that forms a tectonic **Plate**; it contains both **Crust** and part of the underlying **Mantle** and is typically 100 to 200 kilometers thick.

Mafic a type of igneous rock that is relatively rich in magnesium ("ma") and iron ("fic," from *ferric*); basalt is mafic with about 10 percent magnesium oxide and a similar amount of iron oxide.

Magma body of molten rock.

Magnetic field the influence of magnetism on moving electric charges, creating a force that can move these charges; Earth's magnetic field is generated in the core, creating a shield in space that protects the atmosphere from the solar wind.

Mantle portion of Earth between the crust and the core; the upper mantle is composed mainly of the minerals **Olivine** and **Pyroxene**.

Mantle plume a hot upwelling region in the **Mantle** that rises toward Earth's surface, usually triggering volcanism see **Plume**.

Megathrust the giant **Fault** between two tectonic **Plates** in a **Subduction Zone**; prone to some of Earth's largest **Earthquakes**.

Mesozoic meaning "time of middle life"; the **Era** in Earth's history between 250 and 66 million years ago.

Metamorphic rock a rock that has undergone changes from the effects of temperature and pressure after it first formed.

Meteorite a piece of rock or metal that has fallen to Earth's surface from outer space as a meteor; most are **Chondrites** (including carbonaceous chondrites), but some are composed of iron-nickel alloys.

Mica a platy silicate mineral that typically forms crystalline sheets that split along many planes; can be light-colored (muscovite), or dark-colored (biotite).

Microbe a microscopic organism, or microorganism, including bacteria, cyanobacteria (blue-green algae) and other algae.

Mid-ocean ridge a linear zone of shallowing in the middle of oceans; new **Oceanic Crust** is created by volcanic eruptions along its crest through the process of **Seafloor Spreading**. A more general term, without the implication of running down the middle of an ocean, is a seafloor spreading center.

Milankovitch cycles cycles in Earth's climate, fluctuating between cold periods, known as *glacials*, and warm periods known as *interglacials* on timescales of tens to hundreds of thousands of years. These are controlled by small variations in Earth's orbit around the sun, first quantified by the Serbian mathematician Milutin Milanković in the early part of the twentieth century.

Miocene a division of time marking the fourth **Period** in the **Cenozoic Era**, between about 23 and 5 million years ago.

Moho the base of Earth's **Crust**, marking the change in rock types at the boundary with the underlying **Mantle**, first discovered in 1909 from an analysis of **Earthquake** vibrations by the Croatian seismologist Andrija Mohorovičić.

Norite a type of **Gabbro**, but with some differences in the nature of the rock minerals, associated with rich deposits of copper in South Africa.

Oceanic crust the crust beneath the oceans, about 7 kilometers thick, consisting of three principal layers: a thin covering layer of **Sedimentary Rock**; a middle layer of fine-grained **Volcanic Rock (Basalt)**; and a bottom layer of coarse-grained **Igneous Rock** called **Gabbro**.

Oceanic Plateau unusually shallow part of the deep ocean, underlain by thick **Oceanic Crust**, 20 to 30 kilometers thick; thought to have been created by prolific volcanic eruptions on the seafloor above a **Superplume**.

Oceanic trench a long, deep depression in the ocean floor, up to 11 kilometers deep, where a tectonic **Plate** bends downward and sinks back into Earth's interior along a **Megathrust**; the steep sides are prone to **Debris Avalanches**, triggered by **Earthquakes** on the megathrust and often leading to the formation of **Turbidites** that partly fill the trench.

Olivine a silicate mineral rich in magnesium and iron, commonly found in the upper parts of the **Mantle**.

Oncoid a roughly spherical carbonate concretion, usually several centimeters across, made up of internal concentric layers of calcium carbonate, rock fragments, and sometimes the fossilized remains of microbes that grew on the concretion and played a role in the precipitation of carbonate.

Ooid a nearly spherical carbonate concretion, less than 2 millimeters across, with internal concentric layers formed from the precipitation of carbonate from sea water on a small nucleus that subsequently rolls around the sea bed, building up layers as more carbonate precipitates.

Ophiolite remnants of **Oceanic Crust** caught up in a mountain chain during **Subduction** or **Continental Collision**, often forming long, thin belts of rock.

P wave a form of seismic vibration generated by an **Earthquake** that travels through Earth's interior; the motion of the vibrating rock is similar to the compression and extension of a mechanical spring.

Paleocurrent direction of sediment transport carried by water or wind and preserved in sedimentary structures in rocks.

Paleozoic meaning "time of old life"; the **Era** in Earth's history between about 540 and 250 million years ago.

Pangea a supercontinent made up of all the current continents when they fitted together to form a single landmass about 250 million years ago; Pangea subsequently split, leaving behind Gondwana that subsequently fragmented into today's southern continents plus Arabia and India.

Period a subdivision of a geological **Era**, usually lasting from tens of millions to over a hundred million years.

Phanerozoic meaning "visible life"; the **Eon** of Earth's history beginning about 540 million years ago, marked by the appearance of organisms with hard parts such as shells or skeletons.

Photoferrotrophy a biochemical reaction carried out by photoferrotrophs, such as certain types of bacteria, that uses anoxygenic photosynthesis to oxidize iron; photoferrotrophs may have been important in the creation of **Banded Iron Formation** in environments starved of oxygen, particularly in early Earth prior to 2.3 billion years ago.

Pisiod see **Ooid**, but concretions are now more than 2 millimeters across, and typically up to 10 millimeters across.

Plates or tectonic plates the curved, rigid parts of Earth's outer shell (synonymous with **Lithosphere**) that move relative to one another.

Plate tectonics the scientific theory describing the outer part of Earth in terms of large rigid tectonic **Plates** that move relative to each other; phenomena such as **Earthquakes, Seafloor Spreading, Subduction, Continental Collision, Volcanic Arcs**, occur at the edges of the tectonic plates.

Plume an upwelling jet of unusually hot rock in the **Mantle**, often originating near the core-mantle boundary; plumes start to melt when they reach shallower parts of the mantle, causing the eruption of **Basalts** in oceanic islands like Hawaii. In early Earth, plumes may have caused eruptions of **Komatiite**.

Post-glacial rebound the process by which Earth's surface returns to its original shape after being depressed by the weight of vast ice sheets.

Precambrian the portion of Earth's history that predates the Paleozoic **Era**, spanning from Earth's formation about 4.6 billion years ago to roughly 540 million years ago.

Proterozoic the **Eon** of geological time from about 2.5 billion to 540 million years ago.

Pyroxene a silicate mineral commonly found in igneous rocks and in the upper part of Earth's **Mantle**.

Quartz a silicate mineral composed entirely of silicon and oxygen.

Quartzite a sandstone composed almost exclusively of quartz grains.

Radioactivity the phenomenon exhibited by certain unstable elements that spontaneously decay into daughter elements at a predictable rate, forming the basis for nearly all modern methods of dating rocks.

Radiolaria unicellular organisms found as zooplankton throughout the oceans. They extract silica from the water to build their distinctive skeletons, and when they die, the skeletons sink to the deep sea floor, eventually accumulating as layers of **Chert**.

Rayleigh Number a parameter, named after the nineteenth-century British physicist Lord Rayleigh, that combines various physical properties of a fluid to determine whether **Convection** is likely to occur; a value greater than a thousand indicates a likelihood of convection–Earth's **Mantle** has a Rayleigh Number greater than a million.

Red bed(s) sandstone or interbedded sequences of sandstone, siltstone, and shale with a pervasive red color caused by minute quantities of iron oxide and other compounds of iron (such as hematite and goethite); often river-lain sediments that have had prolonged exposure to oxygen.

Rhythmite multi-layered fine-grained sedimentary rock laid down with obvious periodicity or regularity as a result of repetitive sedimentary processes; commonly, it records the back and forth motion of tidal currents, providing a detailed record of Earth's tides back through geological time.

Rodinia an early **Supercontinent** that reached its maximum extent about 900 million years ago, before starting to fragment about 750 million years ago.

Rubidium an element with a radioactive **Isotope** (Rubidium with an atomic mass of 87) that decays to strontium (strontium with an atomic mass of 87); the parent-daughter pair of rubidium and strontium provides the basis of an important method of dating **Igneous Rocks**, and it has been used extensively to date some of the oldest rocks on Earth.

S wave a type of seismic vibration, generated by an **Earthquake**, that travels through Earth's interior; the motion of the vibrating rock is similar to the sideways movement of a snake.

Schist a medium-grained **Metamorphic Rock**, usually containing platy (sheet-like) minerals such as **Mica**, that tends to break along numerous parallel planes.

Seafloor spreading (center) the process by which **Oceanic Crust** is created as tectonic **Plates** move apart at a seafloor spreading center (or **Mid Ocean Ridge**) above an upwelling of hot mantle rock.

Sedimentary rock a rock that has formed through the settling out of material, either in air or water, such as calcium carbonate precipitated from water, or grains of rock or shell fragments that were transported by water currents or wind, and then deposited in layers to form—depending on grain size—mud, siltstone, sandstone or conglomerate.

Serpentinite rock made up of the mineral serpentine, formed by the metamorphism of mantle rocks such as peridotite.

Siderophile an element that is typically found associated with iron, therefore classed as an iron lover (from the Ancient Greek *sideros*, meaning "iron," and *philos*, meaning "love").

Silicate a type of mineral that makes up most of Earth's **Crust** and **Mantle**, composed primarily of silicon and oxygen forming a basic building block of a tetrahedron, with other elements added; the arrangement of these tetrahedra, together with the detailed composition, define different silicate minerals.

Solar nebula the gaseous cloud of matter from which the sun and planets formed by condensation and accretion.

Strata the layers of sedimentary rock; also referred to as bedding.

Stromatolite a calcareous mound built up of layers of limestone due to the activity of a **Microbe** such as cyanobacteria (blue-green algae), together with trapped sediment; forming some of the earliest known fossils going back to the **Archean Eon**, and still growing today in lagoons in western Australia.

Subduction (zone) the process (or region of the Earth) by which a tectonic **Plate** bends down and sinks back into Earth's **Mantle** along a **Megathrust**; typically associated with **Ocean Trenches**, **Earthquakes**, and **Volcanic Arcs**.

Supercontinent an assembly of many smaller continents to make a single much larger continent, sometimes comprising all the planet's landmass (for example, **Pangea**); a result of plate tectonics and multiple continental collisions and forming periodically in Earth's history.

Superplume an unusually large **Plume** in Earth's **Mantle** that produces very large amounts of **Magma**, sometimes erupting very rapidly to build up large underwater oceanic plateaus; superplumes may have played a role in triggering the break-up of continents, mass extinctions of life, and in early Earth, eruptions of **Komatiite**.

Syncline (or Anticline) a *syncline* is a fold in the layers where the **Strata** or **Bedding** typically form a "V" or "U" shape and they become younger *toward* the middle of the structure. An *anticline* is a fold in rock layers where the beds typically dip outward to form an upside-down "V" or "U" shape, with beds becoming younger *away* from the middle of the structure. If the original orientation ("way up") of the beds is unknown, the more general terms *synform*, and *antiform* are used.

Synsedimentary deformation a term describing **Deformation** (faulting and folding) of the bedrock at Earth's surface that is occurring synchronously with sedimentary processes, such as deposition or erosion by rivers, so that newly deposited sedimentary layers are themselves soon folded and faulted, and then are partly eroded away.

Tektite a small millimeter- to centimeter-sized glassy ball, typically formed when terrestrial rock is melted in the intense heat of a meteorite impact and ejected into the atmosphere, where it cools and solidifies as it falls back to Earth.

Thin section a very thin slice of rock (typically 30 micrometers thick) that is translucent, revealing under the microscope the nature of the rock's individual crystals and other features.

Tonalite a light-colored and coarse-grained intrusive **Igneous Rock**, similar to **Granite** but with mainly plagioclase **Feldspar**, with smaller amounts of dark-colored minerals such as **Amphibole**, and is common in the **Cratons**; first described from the Tonale Pass between Austria and Italy.

Triassic division of time that is the first **Period** of the **Mesozoic Era** between about 252 and 201 million years ago when the first dinosaurs appeared on Earth.

Trondhjemite a coarse-grained and light colored intrusive **Igneous Rock**, similar to **Tonalite**, but with very few dark minerals, and common in the **Cratons**; first described from near the city of Trondheim in Norway.

Turbidite a sandstone layer which is the fallout of a turbulent cloud of sand and mud in a submarine **debris avalanche** that typically formed on the steep edges of the oceans, particularly **Oceanic Trenche**s; usually triggered by the violent shaking during an **Earthquake**.

Ultrabasic rock an **Igneous Rock** containing less than 45 percent silica by weight; **Mantle** rocks such as peridotite are typical examples.

Ultramafic a type of **Igneous Rock** that consists mostly of iron- and magnesium-rich silicate minerals, such as **Olivine** and **Pyroxene**, and so is unusually rich in magnesium and iron oxides.

Unconformity the contact between two rock formations that results from substantial erosion before the deposition of the younger formation.

Viscosity a measure of the runniness or stiffness of a fluid, measured in poises; the viscosity of water is about a hundredth of a poise, whereas rocks in Earth's **Mantle**, although still solid, typically have a viscosity over 24 orders of magnitude greater.

Volcanic arc chain of volcanoes that lies above a **Subduction Zone**.

Volcanic rock rock produced as a result of volcanic activity; usually formed when molten **Lava** cools, either as solidified lava flows or as volcanic ash from explosive eruptions.

Way Up (Younging Direction) direction giving the original orientation of a sedimentary layer when it was laid down, pointing to the sky, so that the top and bottom of the layer is defined.

Yellowcake a term use by the mining industry for a powder of uranium oxide; although sometimes yellow, it is typically black or brown.

Zircon a crystal of zirconium silicate that commonly crystalizes in **Acidic** and **Intermediate** igneous rocks, such as **Granite**; contains radioactive uranium which can be used to date the crystallization of individual zircon crystals with very high precision, providing an important method to date geological events in early Earth.

Further Reading

T hroughout this book I have made use of the extensive geological literature on early Earth and the evolution of life and
our planet. This literature is in scientific journals that may
not be easily (or cheaply) accessible to the general reader. And
it is often couched in the jargon of geologists that can make it
opaque to nonspecialists. In my book, I have gone out of my
way to avoid jargon, and to unpack ideas as much as I can so
that they are readily understandable. Most of the ideas and
all the geological facts that I write about can be found in this
literature, but to make the book more readable for a general
audience, I have refrained from the scientific convention of cit-
ing my sources in the main text, or in footnotes. As a result,
many scientists who have played pivotal roles in developing our
understanding of these topics have gone unnamed, but I will
rectify this here with a list of the publications I have found particularly useful, arranged according to topic. As you will see, the
literature is vast, so this is just my own selection and should not
be considered as exhaustive.

When writing specifically about the Barberton Greenstone
Belt, or comparisons with younger rocks in New Zealand, I have

drawn extensively on my own research, and my focus has been on where I have had firsthand experience. There are some scientists I have named in the text, either because I know them or because I think their contribution is so large that it is inappropriate not to name them. But I hope these scientists— and the many others who have worked in southern Africa—will be kind to me and allow me to offer my own interpretations of the rocks, which may differ from theirs. But there is no excuse for getting the facts wrong.

For each topic, I have listed the references in chronological order, rather than the usual alphabetical order, so that one can get a feel for the progression of ideas.

- § -

There are not many accessible overviews of the Barberton Greenstone Belt, both in terms of availability and comprehensibility for a general audience. When I started my research, these were the best that were available, although none of these covered in any detail the part of the greenstone belt where I was working in Eswatini (formerly Swaziland):

Visser, D. J. L. (compiler), O. R. Van Eeden, G. K. Joubert, A. P. G. Sohnge, J. S. van Zyl, P. J. Rossow, et al. 1956. *The Geology of the Barberton Area.* Special Publication Geological Survey of South Africa 15.
Anhaeusser, C. R., ed. 1980. *Barberton Greenstone Belt.* Geological Society of South Africa, Special Publication 9.

Since then, there have been a number of syntheses or compilations of geological research, again focusing on the greenstone belt in South Africa and not Eswatini:

de Ronde, C. E., and M. J. de Wit. 1994. "Tectonic History of the Barberton Greenstone Belt, South Africa: 490 Million Years of Archean Crustal Evolution." *Tectonics* 13, no. 4: 983–1005.
de Wit, M. J., and L. D. Ashwal, eds. 1997. *Greenstone Belts.* Oxford University Press.
Lowe, D. R., and G. R. Byerly, eds. 1999. *Geologic Evolution of the Barberton Greenstone Belt, South Africa.* Geological Society of America.
Johnson, M. R., C. R. Anhaeusser, and R. J. Thomas. 2006. *The Geology of South Africa.* Geological Society of South Africa.
Lowe, D. R., and G. R. Byerly. 2007. "Overview of the Geology of the Barberton Greenstone Belt and Vicinity: Implications for Early Crustal Development." In *Earth's Oldest Rocks*, 2nd ed., ed. Martin J. Van Kranendonk, R. Hugh Smithies, and Vickie C. Bennett. Elsevier.
Tankard, A. J., M. Martin, K. A. Eriksson, D. K. Hobday, D. R. Hunter, and W. E. L. Minter. 2012. *Crustal Evolution of Southern Africa: 3.8 Billion Years of Earth History.* Springer Science & Business Media.

Byerly, G. R., D. R. Lowe, C. Heubeck. 2018. "Geologic Evolution of the Barberton Greenstone Belt: A Unique Record of Crustal Development, Surface Processes, and Early Life 3.55 to 3.20 Ga." In *Earth's Oldest Rocks*, 2nd ed., ed. M. J. Van Kranendonk, V. C. Bennett, and J. E. Hoffmann. Elsevier.

de Wit, M., H. Furnes, S. MacLennan, M. Doucouré, B. Schoene, U. Weckmann, et al. 2018. "Paleoarchean Bedrock Lithologies Across the Makhonjwa Mountains of South Africa and Swaziland Linked to Geochemical, Magnetic and Tectonic Data Reveal Early Plate Tectonic Genes Flanking Subduction Margins." *Geoscience Frontiers* 9, no. 3: 603–65.

Kröner A., and A. Hofmann, eds. 2019. *The Archean Geology of the Kaapvaal Craton, Southern Africa*. Regional Geology Reviews. Springer.

Don Lowe and his colleagues have also published a geological map of the western and central part of the Barberton Greenstone Belt in South Africa:

Lowe, D., G. R. Byerly, and C. Heubeck. 2012. *Geologic Map of the West-Central Barberton Greenstone Belt, South Africa*. 1:25,000 scale. Geological Society of America.

And finally, there is my own PhD thesis:

Lamb, Simon. 1984. *Geology of Part of the Archaean Barberton Greenstone Belt, Swaziland*. Unpublished PhD thesis, Department of Earth Sciences, University of Cambridge, England.

CHAPTER 1

This chapter is a general introduction to my project in Eswatini and the range of geological concepts needed to make sense of it. I have explored many of these ideas in more detail in my previous books:

Lamb, Simon, and David Sington. 1998. *Earth Story—The Forces That Have Shaped Our Planet*. Princeton University Press.

Lamb, Simon. 2004. *Devil in the Mountain: A Search for the Origin of the Andes*. Princeton University Press.

For the discussion on plate tectonics, see the further reading material about plate tectonics given for chapter 7.

I would also strongly recommend the following books, which give excellent overviews of life on early Earth, and its subsequent history:

Knoll, Andrew H. 2003. *Life on a Young Planet: The First Three Billion Years of Evolution on Earth*. Princeton University Press.

Fortey, Richard. 2004. *Earth: An Intimate History*. Knopf.

Arndt, N. T., and E. G. Nisbet. 2012. "Processes on the Young Earth and the Habitats of Early Life." *Annual Review of Earth and Planetary Sciences* 40, no. 1: 521–49.

Nisbet, E. G. 2012. *The Young Earth: An Introduction to Archean Geology*. Springer Science & Business Media.

Bjornerud, Marcia. 2018. *Timefulness: How Thinking Like a Geologist Can Help Save the World*. Princeton University Press.

Knoll, Andrew H. 2021. *A Brief History of the Earth in Eight Chapters*. Harper Collins.

CHAPTER 2

This chapter is a more detailed introduction to the geology of southern Africa, and to field geology techniques. Again, I would point to my previous books *Earth Story* (op. cit.) and *Devil in the Mountain* (op. cit.).

I also refer to some of the early papers on mapping of the Barberton Greenstone Belt, dating the granites in the Kaapvaal Craton, as well as Alexander Macgregor's idea of "gregarious batholiths":

Hall, A. L. 1918. "The Geology of the Barberton Gold Mining District, including adjoining portions of Northern Swaziland." *Memoirs of the Geological Survey of South Africa* 9: 1–324. Government Printing and Stationery Office, South Africa.

Macgregor, A. M. 1951. "Some Milestones in the Precambrian of Southern Rhodesia." *Proceedings of the Geological Society of South Africa* 54: 27–71.

Nicolaysen, L. O., J. W. L. de Villiers, A. J. Burger, and F. W. E. Strelow, 1958. "New Measurements Relating to the Absolute Age of the Transvaal System and of the Bushveld Igneous Complex." *South African Journal of Geology* 61, no. 1: 137–66.

Allsopp, H. L. 1961. "Rb-Sr Age Measurements on Total Rock and Separated-Mineral Fractions from the Old Granite of the Central Transvaal." *Journal of Geophysical Research* 66, no. 5: 1499–1508

Allsopp, H. L., H. R. Roberts, G. D. L. Schreiner, and D. R. Hunter, 1962. "Rb-Sr Age Measurements on Various Swaziland Granites." *Journal of Geophysical Research* 67, no. 13: 5307–13.

A more modern summary of dates, making use of zircon crystal ages is given by:

de Ronde, C. E. J. et al. 1994. "Tectonic History of the Barberton Greenstone Belt, South Africa: 490 million years of Archean Crustal Evolution." *Tectonics* 13, no. 4: 983–1005.

See references in section on Chapter 8 about dating the very ancient rocks in Greenland and Canada.

The pylon nappe that Isabelle and I were shown by Maarten, on our first trip to the Barberton Greenstone Belt, is described in:

de Wit, Maarten. 1982. "Gliding and Overthrust Nappe Tectonics in the Barberton Greenstone Belt." *Journal of Structural Geology* 4, no. 2: 117–36.

I describe the faults and folds on the Eswatini (Swaziland) side of the Makhonjwa Mountains, in the Malolotja Nature Reserve, in:

Lamb, Simon. 1984. "Structures on the Eastern Margin of the Archean Barberton Greenstone Belt, Northwest Swaziland." In *Precambrian Tectonics Illustrated*, ed. A. Kroner and R. Greiling. E. Schweizerbart'sche Verlagsbuchhandlung.

CHAPTER 3

This chapter deals with the volcanic successions preserved in the Barberton Greenstone Belt, forming the Onverwacht Group. I start with the pioneering work of Morris and Richard Viljoen, and their discovery of komatiites:

Viljoen, M. J. and R. P. Viljoen. 1969. "The Geology and Geochemistry of the Lower Ultra-mafic Unit of the Onverwacht Group and a Proposed New Class of Igneous Rocks." In *Upper Mantle Project*, vol. 2. *Special publication of the Geological Society of South Africa*. Geological Society of South Africa.

Research into these rocks by Maarten de Wit, Harald Furnes, and their colleagues can be found in this series of important papers, starting in 1987:

de Wit, M. J., R. A. Hart, and R. J. Hart. 1987. "The Jamestown Ophiolite Complex, Barberton Mountain Belt: A Section Through 3.5 Ga Oceanic Crust." *Journal of African Earth Sciences* 6, no. 5: 681–730.
de Wit, M. J., R. Armstrong, R. J. Hart, and A. H. Wilson. 1987. "Felsic Igneous Rocks Within the 3.3-to 3.5-Ga Barberton Greenstone Belt: High Crustal Level Equivalents of the Surrounding Tonalite-Trondhjemite Terrain, Emplaced During Thrusting." *Tectonics* 6, no. 5: 529–49.
Furnes, H., M. J. de Wit, B. Robins, and N. R. Sandstå. 2011. "Volcanic Evolution of the Upper Onverwacht Suite, Barberton Greenstone Belt, South Africa." *Precambrian Research* 186, nos. 1–4: 28–50.
Furnes, H., M. J. de Wit, and B. Robins. 2013. "A Review of New Interpretations of the Tectonostratigraphy, Geochemistry, and Evolution of the Onverwacht Suite, Barberton Greenstone Belt, South Africa." *Gondwana Research* 23, no. 2: 403–28.

Jesse Dann's elegant reevaluation of whether the komatiites were intrusive or extrusive rocks can be found in:

Dann, J. C. and T. L. Grove. 2007. "Volcanology of the Barberton Greenstone Belt, South Africa: Inflation and Evolution of Flow Fields." *Developments in Precambrian Geology* 15: 527–70.

There is a large body of scientific literature on the origin of komatiites and what they may tell us about processes in the mantle. I find the following to be particularly helpful:

Nisbet, E. G., M. J. Cheadle, N. T. Arndt, and M. J. Bickle. 1993. "Constraining the Potential Temperature of the Archean Mantle: A Review of the Evidence from Komatiites." *Lithos* 30, nos. 3–4: 291–307.

Parman, S. W., J. C. Dann, T. L. Grove, and M. J. de Wit. 1997. "Emplacement Conditions of Komatiite Magmas from the 3.49 Ga Komati Formation, Barberton Greenstone Belt, South Africa." *Earth and Planetary Science Letters* 150, nos. 3–4: 303–23.

Parman, S. W., T. L. Grove, J. C. Dann, and M. J. de Wit. 2004. "A Subduction Origin for Komatiites and Cratonic Lithospheric Mantle." *South African Journal of Geology* 107, nos. 1–2: 107–18.

Grove, T. L. and S. W. Parman. 2004. "Thermal Evolution of the Earth as Recorded by Komatiites." *Earth and Planetary Science Letters* 219, nos. 3–4: 173–87.

Sobolev, A. V., E. V. Asafov, A. A. Gurenko, N. T. Arndt, V. G. Batanova, M. V. Portnyagin, et al. 2019. "Deep Hydrous Mantle Reservoir Provides Evidence for Crustal Recycling Before 3.3 Billion Years Ago." *Nature* 571, no. 7766: 555–59.

McKenzie, D. 2020. "Speculations on the Generation and Movement of Komatiites." *Journal of Petrology* 61, no. 7: egaa061, https://doi.org/10.1093/petrology/egaa061.

Harald Furnes's work with Maarten de Wit and colleagues on sheeted dikes in the 3.7–3.8 billion-year-old Isua rocks is described in:

Furnes, H., M. J. de Wit, H. Staudigel, M. Rosing, and K. Muehlenbachs. 2007. "A Vestige of Earth's Oldest Ophiolite." *Science* 315, no. 5819: 1704–07.

Furnes, H., M. Rosing, Y. Dilek, and M. J. de Wit. 2009. "Isua Supracrustal Belt (Greenland)—A Vestige of a 3.8 Ga Suprasubduction Zone Ophiolite, and the Implications for Archean Geology." *Lithos* 113, nos. 1–2: 115–32.

There are many papers on the origin of the cherts in the Barberton Greenstone Belt. Here are a selection that I have found useful when writing about the Onverwacht Group:

Lowe, D. R., and L. P. Knauth. 1977. "Sedimentology of the Onverwacht Group (3.4 Billion Years), Transvaal, South Africa, and Its Bearing on the Characteristics and volution of the Early Earth." *The Journal of Geology* 85, no. 6: 699–723.

Stanistreet, I. G., M. J. de Wit, and R. E. Fripp. 1981. "Do Graded Units of Accretionary Spheroids in the Barberton Greenstone Belt Indicate Archean Deep Water Environment?" *Nature* 293, no. 5830: 280–84.

de Wit, M. J., R. Hart, A. Martin, and P. Abbott. 1982. "Archean Abiogenic and Probable Biogenic Structures Associated with Mineralized Hydrothermal Vent Systems and Regional Metasomatism, with Implications for Greenstone Belt Studies." *Economic Geology* 77, no. 8: 1783–1802.

Paris, I., I. G. Stanistreet, and M. J. Hughes. 1985. "Cherts of the Barberton Greenstone Belt Interpreted as Products of Submarine Exhalative Activity." *The Journal of Geology* 93, no. 2: 111–29.

de Ronde, C. E., M. J. de Wit, and E. T. Spooner. 1994. "Early Archean (> 3.2 Ga) Fe-oxide-rich, Hydrothermal Discharge Vents in the Barberton Greenstone Belt, South Africa." *Geological Society of America Bulletin* 106, no. 1: 86–104.

Knauth, L. P., and D. R. Lowe. 2003. "High Archean Climatic Temperature Inferred from Oxygen Isotope Geochemistry of Cherts in the 3.5 Ga Swaziland Supergroup, South Africa." *Geological Society of America Bulletin* 115, no. 5: 566–80.

Hofmann, A., and C. Harris. 2008. "Silica Alteration Zones in the Barberton Greenstone Belt: A Window into Subseafloor Processes 3.5–3.3 Ga Ago." *Chemical Geology* 257, nos. 3–4: 221–39.

de Wit, M. J., and H. Furnes. 2016. "3.5-Ga Hydrothermal Fields and Diamictites in the Barberton Greenstone Belt—Paleoarchean Crust in Cold Environments." *Science Advances* 2, no. 2: e1500368. https://doi.org/10.1126/sciadv.1500368.

Evidence for early life in the Onverwacht and Fig Tree Groups is described in:

Barghoorn, E. S. and J. W. Schopf. 1966. "Microorganisms Three Billion Years Old from the Precambrian of South Africa." *Science* 152, no. 3723: 758–63.

Schopf, J. W. and E. S. Barghoorn. 1967. "Alga-Like Fossils from the Early Precambrian of South Africa." *Science* 156, no. 3774: 508–12.

Engel, A. E., B. Nagy, L. A. Nagy, C. G. Engel, C. O. Kremp, and C. M. Drew. 1968. "Alga-Like Forms in Onverwacht Series, South Africa: Oldest Recognized Lifelike Forms on Earth." *Science* 161, no. 3845: 1005–1008.

Knoll, A. H. and E. S. Barghoorn. 1977. "Archean Microfossils Showing Cell Division from the Swaziland System of South Africa." *Science* 198, no. 4315: 396–98.

Walsh, M. W. 1992. "Microfossils and Possible Microfossils from the Early Archean Onverwacht Group, Barberton Mountain Land, South Africa." *Precambrian Research* 54, nos. 2–4: 271–93.

Fliegel, D., J. Kosler, N. McLoughlin, A. Simonetti, M. J. De Wit, R. Wirth, et al. 2010. "In-Situ Dating of the Earth's Oldest Trace Fossil at 3.34 Ga." *Earth and Planetary Science Letters* 299, nos. 3–4: 290–98.

Frances Westall and colleagues have published a number of papers on fossil microorganisms preserved in the cherts of the Onverwacht Group:

Westall, F., M. J. de Wit, J. Dann, S. van der Gaast, C. E. de Ronde, and D. Gerneke. 2001. "Early Archean Fossil Bacteria and Biofilms in Hydrothermally-Influenced Sediments from the Barberton Greenstone Belt, South Africa." *Precambrian Research* 106, nos. 1–2: 93–116.

Westall, F., B. Cavalazzi, L. Lemelle, Y. Marrocchi, J. N. Rouzaud, A. Simionovici, et al. 2011. "Implications of In Situ Calcification for Photosynthesis in a ~3.3 Ga-old Microbial

Biofilm from the Barberton Greenstone Belt, South Africa." *Earth and Planetary Science Letters* 310, nos. 3–4: 468–79.

Westall, F., K. A. Campbell, J. G. Bréhéret, F. Foucher, P. Gautret, A. Hubert, et al. 2015. "Archean (3.33 Ga) Microbe-Sediment Systems Were Diverse and Flourished in a Hydrothermal Context." *Geology* 43, no. 7: 615–18.

CHAPTER 4

This chapter examines the sedimentary sequence in the Barberton Greenstone Belt that formed in deep to shallow water on the edge of land, comprising the top part of the Onverwacht Group, together with the Fig Tree Group. You can find more detailed descriptions of the rocks I talk about in my paper with Isabelle Paris:

Lamb S. H., and I. Paris. 1988. "The Post-Onverwacht Group Stratigraphy in the SE Part of the Archean Barberton Greenstone Belt." *Journal of African Earth Sciences* 7: 285–306.

And also:

Nocita, B. W., and D. R. Lowe. 1990. "Fan-Delta Sequence in the Archean Fig Tree Group, Barberton Greenstone Belt, South Africa. *Precambrian Research* 48, no. 4: 375–93.

Lowe, D. R., and G. R. Byerly, eds. 1999. *Geologic Evolution of the Barberton Greenstone Belt, South Africa*. Geological Society of America Special Papers 329. Geological Society of America.

Trower, E. J., and D. R. Lowe. 2016. "Sedimentology of the~ 3.3 Ga Upper Mendon Formation, Barberton Greenstone Belt, South Africa." *Precambrian Research* 281: 473–94.

Drabon, N., A. Galić, P. R. D. Mason, D. R. Lowe. 2019. "Provenance and Tectonic Implications of the 3.28–3.23 Ga Fig Tree Group, Central Barberton Greenstone Belt, South Africa." *Precambrian Research* 325: 1–19.

Lowe, D. R., and G. R. Byerly. 2024. "Geology of the Eastern Barberton Greenstone Belt, South Africa: Early Deformation and the Role of Large Meteor Impacts." *American Journal of Science* 324, no. 12: article 122938. https://doi.org/10.2475/001c.122938.

See chapter 3 for references to descriptions of cherts and their many varieties. The mud pool structures and possible stromatolites are described by Maarten de Wit in:

de Wit, M. J., R. Hart, A. Martin, and P. Abbott. 1982. "Archean Abiogenic and Probable Biogenic Structures Associated with Mineralized Hydrothermal Vent Systems and Regional Metasomatism, with Implications for Greenstone Belt Studies." *Economic Geology* 77, no. 8: 1783–1802.

de Ronde, C. E., M. J. de Wit, and E. T. Spooner. 1994. "Early Archean (> 3.2 Ga) Fe-Oxide-Rich, Hydrothermal Discharge Vents in the Barberton Greenstone Belt, South Africa." *Geological Society of America Bulletin* 106, no. 1: 86–104.

See also:

Hofmann, A. 2005. "The Geochemistry of Sedimentary Rocks from the Fig Tree Group, Barberton Greenstone Belt: Implications for Tectonic, Hydrothermal and Surface Processes During Mid-Archean Times." *Precambrian Research* 143, nos. 1–4: 23–49.

Later papers on these topics are given in Frances Westall's papers, referred to in chapter 3; see also:

Byerly, G. R., D. R. Lowe, and M. M. Walsh. 1986. "Stromatolites from the 3,300–3,500-Myr Swaziland Supergroup, Barberton Mountain Land, South Africa." *Nature* 319, no. 6053: 489–91.

The research on giant submarine landslides, debris avalanches, and turbidites, including their discovery in the modern oceans and Miocene examples from the Great Marlborough Conglomerate, New Zealand, can be found in:

King, L. C. 1937. "The Tertiary Sequence in North-Eastern Marlborough." *Transactions and Proceedings of the Royal Society of New Zealand* 67: 21–32.
Heezen, B. C., and W. M. Ewing. 1952. "Turbidity Currents and Submarine Slumps, and the 1929 Grand Banks [Newfoundland] Earthquake." *American Journal of Science* 250, no. 12: 849–73.
Heezen, B. C., D. B. Ericson, and M. Ewing. 1954. "Further Evidence for a Turbidity Current Following the 1929 Grand Banks Earthquake." *Deep Sea Research* 1, no. 4: 193–202.
Lamb, S., and H. M. Bibby. 1989. "The Last 25 Ma of Rotational Deformation in Part of the New Zealand Plate-Boundary Zone." *Journal of Structural Geology* 11: 473–92.
Mountjoy, J. J., J. D. Howarth, A. R. Orpin, P. M. Barnes, D. A. Bowden, A. A. Rowden, et al. 2018. "Earthquakes Drive Large-Scale Submarine Canyon Development and Sediment Supply to Deep-Ocean Basins." *Science Advances* 4, no. 3: 3748. https://doi.org/10.1126/sciadv.aar3748.
Lamb, S., and C. E. de Ronde. 2024. "Large-Scale Submarine Landslides in the Barberton Greenstone Belt, Southern Africa—Evidence for Subduction and Great Earthquakes in the Paleoarchean." *Geology* 52, no. 6: 390–94.

Cornel de Ronde's new map of part of the Barberton Greenstone Belt, showing evidence for large slide blocks in the Onverwacht and Fig Tree Groups, is published as:

de Ronde C. E. J. 2021. *Geological Map of the Central and Western Parts of the Barberton Greenstone Belt, Makhonjwa Mountains.* GNS Science. 1 sheet, scale 1:14,100. DOI:10.21420/HDAK-TP07.

Details on the fossil meteorite impact fallout, cracking of cherts, and barite deposits, can found in:

Reimer, T. O. 1980. "Archean Sedimentary Barite Deposits of the Swaziland Supergroup (Barberton Mountain Land, South Africa)." *Precambrian Research* 12.

Lowe, D. R. 2013. "Crustal Fracturing and Chert Dike Formation Triggered by Large Meteorite Impacts, ca. 3.260 Ga, Barberton Greenstone Belt, South Africa." *Geological Society of America Bulletin* 125, nos. 5–6: 894–912.

Lowe, D. R., N. Drabon, and G. R. Byerly. 2019. "Crustal Fracturing, Unconformities, and Barite Deposition, 3.26–3.23 Ga, Barberton Greenstone Belt, South Africa." *Precambrian Research* 327: 34–46.

Some of the original work on the evidence for continental crust formation during and after the deposition of the Fig Tree Group sediments can be found in:

de Wit, M. J., R. Armstrong, R. J. Hart, and A. H. Wilson. 1987. "Felsic Igneous Rocks within the 3.3-to 3.5-Ga Barberton Greenstone Belt: High Crustal Level Equivalents of the Surrounding Tonalite-Trondhjemite Terrain, Emplaced During Thrusting." *Tectonics* 6, no. 5: 529–49.

de Ronde, C. E. J., S. Kamo, D. W. Davis, M. J. de Wit, and E. T. C. Spooner. 1991. "Field, Geochemical and U-Pb Isotopic Constraints from Hypabyssal Felsic Intrusions within the Barberton Greenstone Belt, South Africa: Implications for Tectonics and the Timing of Gold Mineralization." *Precambrian Research* 49, nos. 3–4: 261–80.

Lowe, D. R., and G. R. Byerly, eds. 1999. *Geologic Evolution of the Barberton Greenstone Belt, South Africa*. Geological Society of America Special Paper 329. Geological Society of America.

CHAPTER 5

This chapter deals with continental collision in Earth's early history, as revealed in the Barberton Greenstone Belt by the folding and faulting of rock layers, and the evidence preserved in the Moodies Group for rivers and tidal currents in shallow seas. These rocks, where they are exposed in Eswatini (Swaziland), are described in my paper with Isabelle Paris:

Lamb, S. H., and I. Paris. 1988. "The Post-Onverwacht Group Stratigraphy in the SE Part of the Archean Barberton Greenstone Belt." *Journal of African Earth Sciences* 7: 285–306.

You can find more regional descriptions of the Moodies Group in South Africa in:

Eriksson, K. A. 1979. "Marginal Marine Depositional Processes from the Archean Moodies Group, Barberton Mountain Land, South Africa: Evidence and Significance." *Precambrian Research* 8, nos. 3–4: 153–82.

Eriksson, K. A., and E. L. Simpson. 2000. "Quantifying the Oldest Tidal Record: the 3.2 Ga Moodies Group, Barberton Greenstone Belt, South Africa." *Geology* 28, no. 9: 831–34.

Heubeck, C. 2019. "The Moodies Group—A High-Resolution Archive of Archean Surface Processes and Basin-Forming Mechanisms." In *The Archean Geology of the Kaapvaal Craton, Southern Africa*, ed. A. Kröner and A. Hofmann. *Regional Geology Reviews*. Springer.

I describe both the fold at Ngwenya, in my study area in the Malolotja Nature Reserve, Eswatini (Swaziland), and also the remarkably similar one in New Zealand, in:

Lamb, S. H., and P. Vella. 1987. "The Last Million Years of Deformation in Part of the New Zealand Plate-Boundary Zone. *Journal of Structural Geology* 9: 877–91.

Lamb, S. H. 1987. "Synsedimentary Tectonic Deformation in the Archean Barberton Greenstone Belt—A Comparison with the Quaternary." *Geology* 15: 565–68.

I also describe the great faults that shuffle the ancient sequences in my Eswatini study area, in:

Lamb, S. H. 1984. "Structures on the Eastern Margin of the Archean Barberton Greenstone Belt, Northwest Swaziland." In *Precambrian Tectonics Illustrated*, ed. A. Kröner and R. Greiling. E. Schweizerbart'sche Verlagsbuchhandlung.

For a description of deformation in the South African part of the Barberton Greenstone Belt, see:

Heubeck, C., and D. R. Lowe. 1994. "Late Syndepositional Deformation and Detachment Tectonics in the Barberton Greenstone Belt, South Africa." *Tectonics* 13, no. 6: 1514–36.

I consulted the following original papers on the concept of a lighter crust floating on a denser underling mantle, as well as nineteenth-century perspectives on the nature of Earth:

Airy, G. B. 1855. "On the Computation of the Effect of the Attraction of Mountain-Masses, as Disturbing the Apparent Astronomical Latitude of Stations in Geodetic Surveys." *Philosophical Transactions of the Royal Society of London* 145: 101–104.

Airy, G. B. 1856. "XIV. Account of Pendulum Experiments Undertaken in the Harton Colliery, for the Purpose of Determining the Mean Density of the Earth. & XV. Supplement to the 'Account of pendulum experiments undertaken in the Harton Colliery'; Being an Account of Experiments Undertaken to Determine the Correction for the Temperature of the Pendulum." *Philosophical Transactions of the Royal Society of London* 146: 297–355.

Herschel, Sir John F. W. 1861. *Physical Geography: From the Encyclopaedia Britannica.* 1st ed. Adam and Charles Black.

If you are interested in how the thickness of the tectonic plates affects surface elevations in the continents, see:

Lamb, S. H., J. D. Moore, M. Perez-Gussinye, and T. Stern. 2020. "Global Whole Lithosphere Isostasy: Implications for Surface Elevations, Structure, Strength, and Densities of the Continental Lithosphere." *Geochemistry, Geophysics, Geosystems* 21, no. 10: e2020GC009150. https://doi.org/10.1029/2020GC009150.

For overviews of the origin of the continents, see:

De Wit, M. J., C. E. de Ronde, M. Tredoux, C. Roering, R. J. Hart, R. A. Armstrong, et al. 1992. "Formation of an Archean Continent." *Nature* 357, no. 6379: 553–62.

de Wit, M. J., and R. A. Hart. 1993. "Earth's Earliest Continental Lithosphere, Hydrothermal Flux and Crustal Recycling." *Lithos* 30, nos. 3–4: 309–35.

Rudnick, R. L. 1995. "Making Continental Crust." *Nature* 378, no. 6557: 571–78.

Moyen, J. F., and H. Martin. 2012. "Forty Years of TTG Research." *Lithos* 148: 312–36.

Palin, R. M., R. W. White, and E. C. Green. 2016. "Partial Melting of Metabasic Rocks and the Generation of Tonalitic–Trondhjemitic–Granodioritic (TTG) Crust in the Archean: Constraints from Phase Equilibrium Modelling." *Precambrian Research* 287: 73–90.

Nebel, O, F. A. Capitanio, J.-F. Moyen, R. F. Weinberg, F. Clos, Y. J. Nebel-Jacobsen, et al. 2018. "When Crust Comes of Age: On the Chemical Evolution of Archean, Felsic Continental Crust by Crustal Drip Tectonics." *Philosophical Transactions of the Royal Society A: Mathematical, Physical and Engineering Sciences* 376: 20180103. http://dx.doi.org/10.1098/rsta.2018.0103.

Kusky, T., B. F. Windley, A. Polat, L. Wang, W. Ning, and Y. Zhong. 2021. "Archean Dome-and-Basin Style Structures Form During Growth and Death of Intraoceanic and Continental Margin Arcs in Accretionary Orogens." *Earth-Science Reviews* 220: 103725. https://doi.org/10.1016/j.earscirev.2021.103725.

I used the following papers for ideas about the final closing-up of an ocean and continental collision, which terminated the life of the Barberton Greenstone Belt, with the rise of granitic diapirs, leading to the creation of the Kaapvaal Craton, and the timetable for these events:

de Wit, M. J., R. Armstrong, R. J. Hart, and A H. Wilson. 1987. "Felsic Igneous Rocks within the 3.3-to 3.5-Ga Barberton Greenstone Belt: High Crustal Level Equivalents of the Surrounding Tonalite-Trondhjemite Terrain, Emplaced During Thrusting." *Tectonics* 6, no. 5; 529–49.

de Ronde, C. E. J., S. Kamo, D. W. Davis, M. J. de Wit, and E. T. C. Spooner. 1991. "Field, Geochemical and U-Pb Isotopic Constraints from Hypabyssal Felsic Intrusions within the Barberton Greenstone Belt, South Africa: Implications for Tectonics and the Timing of Gold Mineralization." *Precambrian Research* 49, nos. 3–4: 261–80.

Lowe, D. R. 1994. "Accretionary History of the Archean Barberton Greenstone Belt (3.55–3.22 Ga), Southern Africa." *Geology* 22, no. 12: 1099–1102.

Lowe, D. R., and G. R. Byerly, eds. 1999. *Geologic Evolution of the Barberton Greenstone Belt, South Africa*. Geological Society of America Special Paper 329. Geological Society of America

De Ronde, C. E., and S. L. Kamo. 2000. "An Archean Arc-Arc Ollisional Event: A Short-Lived (ca. 3 Myr) Episode, Weltevreden Area, Barberton Greenstone Belt, South Africa." *Journal of African Earth Sciences* 30, no. 2: 219–48.

Robb, L. J., G. Brandl, C. R. Anhaeusser, M. Poujol, M. R. Johnson, and R. J. Thomas. 2006. "Archean Granitoid Intrusions." In *The Geology of South Africa*, ed. M. R. Johnson, C. R. Anhaeusser, and R. J. Thomas. Geological Society of South Africa and Council for Geoscience.

Schoene, B., and S. A. Bowring. 2007. "Determining Accurate Temperature–Time Paths from U-Pb Thermochronology: An Example from the Kaapvaal Craton, Southern Africa." *Geochimica et Cosmochimica Acta* 71, no. 1: 165–85.

Schoene, B., M. J. de Wit, and S. A. Bowring. 2008. "Mesoarchean Assembly and Stabilization of the Eastern Kaapvaal Craton: A Structural-Thermochronological Perspective." *Tectonics* 27, no. 5: TC5010, doi:10.1029/2008TC002267.

He, B., Y. G. Xu, and S. Paterson. 2009. "Magmatic Diapirism of the Fangshan Pluton, Southwest of Beijing, China." *Journal of Structural Geology* 31, no. 6: 615–26.

Schoene, B., and S. A. Bowring. 2010. "Rates and Mechanisms of Mesoarchean Magmatic Arc Construction, Eastern Kaapvaal Craton, Swaziland." *Geological Society of America Bulletin* 122, nos. 3–4: 408–29.

Lana, C., I. Buick, G. Stevens, R. Rossouw, and W. de Wet. 2011. "3230–3200 Ma Post-Orogenic Extension and Mid-Crustal Magmatism Along the Southeastern Margin of the Barberton Greenstone Belt, South Africa." *Journal of Structural Geology* 33, no. 5: 844–58.

Heubeck, C., J. Engelhardt, G. R. Byerly, A. Zeh, B. Sell, T. Luber, et al. 2013. "Timing of Deposition and Deformation of the Moodies Group (Barberton Greenstone Belt, South Africa): Very-High-Resolution of Archean Surface Processes." *Precambrian Research* 231: 236–62.

For a non-plate-tectonic view of the Barberton Greenstone belt, and the link with the surrounding granites, see the following paper:

Van Kranendonk, M. J. 2021. "Gliding and Overthrust Nappe Tectonics of the Barberton Greenstone Belt Revisited: A Review of Deformation Styles and Processes." *South African Journal of Geology* 124, no. 1: 181–210.

CHAPTER 6

This chapter is an overview of some of the major mineral deposits in southern Africa, which I visited in 1982. I have mainly relied on what mine geologists told me about their mines. The nearest I can get to accessible overviews of these deposits are:

Johnson, M. R., C. R. Anhauesser, and R. J. Thomas. 2006. *The Geology of South Africa*. Geological Society of South Africa.

Vijoen, R. P., and M. G. C. Wilson, eds. 2016. *The Great Mineral Fields of Africa*. Special Issue, *Episodes, Journal of International Geoscience*, IUGS 39, no. 2. https://doi.org/10.18814/epiiugs/2016/v39i2/.

Here are a few publications on specific mineral deposits:

Beukes, N. J., J. Gutzmer, and J. Mukhopadhyay. 2003. "The Geology and Genesis of High-Grade Hematite Iron Ore Deposits." *Applied Earth Science* 112, no. 1: 18–25.

Kinnaird, J. A., and P. A. M. Nex. 2007. "A Review of Geological Controls on Uranium Mineralisation in Sheeted Leucogranites Within the Damara Orogen, Namibia." *Applied Earth Science* 116, no. 2: 68–85.

Pretorius, D. A., and K. H. Wolf. 2012. "The Nature of the Witwatersrand Gold-Uranium Deposits." *Handbook of Stratabound and Stratiform Ore Deposits* 7: 29–88.

Clifford, T. N., and E. S. Barton. 2012. "The O'okiep Copper District, Namaqualand, South Africa: A Review of the Geology with Emphasis on the Petrogenesis of the Cupriferous Koperberg Suite." *Mineralium Deposita* 47, no. 8: 837–57.

CHAPTERS 7 AND 8

These chapters deal with the larger picture of the evolution of Earth and life, from its creation to today, placing the Barberton Greenstone Belt in this context. Again, *Earth Story*—both the BBC documentary series and its companion books—covers much of this material in a way designed for a general audience:

Earth Story, BBC Documentary Series. 1998. Series Producer, David Sington and Executive Producer, Richard Reisz. First broadcast in the United Kingdom, November 1–December 27, 1998. BBC.

Lamb, Simon, and David Sington. 1998. *Earth Story: The Shaping of Our World*. Illustrated edition. BBC Books.

Lamb, Simon, and David Sington. 2003. *Earth Story: The Forces That Have Shaped Our Planet*. Princeton University Press.

For those interested in the Gaia hypothesis and the evolution of life, I recommend:

Lovelock, James. 1979. *Gaia: A New Look at Life on Earth*. Oxford University Press.

Lovelock, J., 2000. *The Ages of Gaia: A Biography of Our Living Earth*. Oxford University Press.

Knoll, A. H. 2015. *Life on a Young Planet: The First Three Billion Years of Evolution on Earth*. Princeton University Press.

Also, the following are useful overviews of the evolution of Earth as a habitable planet in comparison to our neighbors Venus and Mars:

Rampino, M. R., and K. Caldeira. 1994. "The Goldilocks Problem: Climatic Evolution and Long-Term Habitability of Terrestrial Planets." *Annual Review of Astronomy and Astrophysics* 32: 83–114.

Kasting, J. F. 2019. "The Goldilocks Planet? How Silicate Weathering Maintains Earth 'Just Right.'" *Elements: An International Magazine of Mineralogy, Geochemistry, and Petrology* 15, no. 4: 235–40.

There is a large body of technical literature in geochemistry that explores the origins of Earth and the moon, as well as Earth's core, magnetic field, atmosphere, oceans, and life. Much of it is highly specialized and difficult to follow without a strong background in the subject. But I have found the following helpful:

Lee, D. C., A. N. Halliday, G. A. Snyder, and L. A. Taylor. 1997. "Age and Origin of the Moon." *Science* 278, no. 5340: 1098–103.

Halliday, A. N., and D. C. Lee. 1999. "Tungsten Isotopes and the Early Development of the Earth and Moon." *Geochimica et Cosmochimica Acta* 63, nos. 23–24: 4157–79.

Eriksson, K. A., and E. L. Simpson. 2000. "Quantifying the Oldest Tidal Record: the 3.2 Ga Moodies Group, Barberton Greenstone Belt, South Africa." *Geology* 28, no. 9: 831–34.

Williams, G. E. 2000. "Geological Constraints on the Precambrian History of Earth's Rotation and the Moon's Orbit." *Reviews of Geophysics* 38, no. 1: 37–59.

Knauth, L. P., and D. R. Lowe. 2003. "High Archean Climatic Temperature Inferred from Oxygen Isotope Geochemistry of Cherts in the 3.5 Ga Swaziland Supergroup, South Africa." *Geological Society of America Bulletin* 115, no. 5: 566–80.

Jacobsen, S. B. 2005. "The Hf-W Isotopic System and the Origin of the Earth and Moon." *Annual Review of Earth and Planetary Sciences* 33: 531–70.

Buick, R. 2008. "When Did Oxygenic Photosynthesis Evolve?" *Philosophical Transactions of the Royal Society B: Biological Sciences* 363, no. 1504: 2731–43.

Bouvier, A., and M. Wadhwa. 2010. "The Age of the Solar System Redefined by the Oldest Pb–Pb Age of a Meteoritic Inclusion." *Nature Geoscience* 3, no. 9: 637–41.

Rosing, M. T., D. K. Bird, N. H. Sleep, and C. J. Bjerrum. 2010. "No Climate Paradox Under the Faint Early Sun." *Nature* 464, no. 7289: 744–47.

Crowe, S. A., L. N. Døssing, N. J. Beukes, M. Bau, S. J. Kruger, R. Frei, et al. 2013. "Atmospheric Oxygenation Three Billion Years Ago." *Nature* 501, no. 7468: 535–38.

Avice, G., and B. Marty. 2014. "The Iodine–Plutonium–Xenon Age of the Moon–Earth System Revisited." *Philosophical Transactions of the Royal Society A: Mathematical, Physical and Engineering Sciences* 372, no. 2024: 20130260, https://doi.org/10.1098/rsta.2013.0260.

Hallis, L. J., G. R. Huss, K. Nagashima, G. J. Taylor, S. A. Halldórsson, D. R. Hilton, M. J. Mottl, and K. J. Meech. 2015. "Evidence for Primordial Water in Earth's Deep Mantle." *Science* 350, no. 6262: 795–97.

Bell, E.A., P. Boehnke, T. M. Harrison, and W. L. Mao. 2015. "Potentially Biogenic Carbon Preserved in a 4.1 Billion-Year-Old Zircon." *Proceedings of the National Academy of Sciences* 112, no. 47: 14518–21.

Tarduno, J. A., R. D. Cottrell, W. J. Davis, F. Nimmo, and R. K. Bono. 2015. "A Hadean to Paleo-archean Geodynamo Recorded by Single Zircon Crystals." *Science* 349, no. 6247: 521–24.

Smith, A. J. 2015. "RESEARCH FOCUS: The Life and Times of Banded Iron Formations." *Geology* 43, no. 12: 1111–12.

Boehnke, P., and T. M. Harrison. 2016. "Illusory Late Heavy Bombardments." *Proceedings of the National Academy of Sciences* 113, no. 39: 10802–06.

Bottke, W. F., and M. D. Norman. 2017. "The Late Heavy Bombardment." *Annual Review of Earth and Planetary Sciences* 45: 619–47.

Ono, S. 2017. "Photochemistry of Sulfur Dioxide and the Origin of Mass-Independent Isotope Fractionation in Earth's Atmosphere." *Annual Review of Earth and Planetary Sciences* 45: 301–29.

Ozaki, K., K. J. Thompson, R. L. Simister, S. A. Crowe, and C. T. Reinhard. 2019. "Anoxygenic Photosynthesis and the Delayed Oxygenation of Earth's Atmosphere." *Nature Communications* 10, no. 1: 1–10.

Thompson, K. J., P. A. Kenward, K. W. Bauer, T. Warchola, T. Gauger, R. Martinez, et al. 2019. "Photoferrotrophy, Deposition of Banded Iron Formations, and Methane Production in Archean Oceans. *Science Advances* 5, no. 11: eaav2869. https://doi.org/10.1126/sciadv.aav2869.

Johnson, B. W., and B. A. Wing. 2020. "Limited Archean Continental Emergence Reflected in an Early Archean 18 O-Enriched Ocean." *Nature Geoscience* 13, no. 3: 243–48.

Catling, D. C., and K. J. Zahnle. 2020. "The Archean Atmosphere." *Science Advances* 6, no. 9: eaax1420. https://doi.org/10.1126/sciadv.aax1420.

Tarduno, J. A., R. D. Cottrell, R. K. Bono, H. Oda, W. J. Davis, M. Fayek, et al. 2020. "Paleo-magnetism Indicates That Primary Magnetite in Zircon Records a Strong Hadean Geodynamo." *Proceedings of the National Academy of Sciences*, 117, no. 5: 2309–18.

Popall R. M., H. Bolhuis, G. Muyzer, and M. Sánchez-Román. 2020. "Stromatolites as Biosignatures of Atmospheric Oxygenation: Carbonate Biomineralization and UV-C Resilience in a Geitlerinema sp.—Dominated Culture." *Frontiers in Microbiology* 11, 948.

Dong, J., R. Fischer, L. Stixrude, and C. Lithgow-Bertelloni. 2021. "Constraining the Volume of Earth's Early Oceans with a Temperature-Dependent Mantle Water Storage Capacity Model." *AGU Advances* 2, no. 1: e2020AV000323. https://doi.org/10.1029/2020AV000323.

The following is some of the original research on dating and understanding the oldest relics of the Earth's surface, preserved at Isua in Greenland and Canada:

Moorbath, S., R. K. O'nions, and R. J. Pankhurst. 1973. "Early Archean Age for the Isua Iron Formation, West Greenland." *Nature* 245, no. 5421: 138–39.

Moorbath, S., R. K. O'nions, and R. J. Pankhurst. 1975. "The Evolution of Early Precambrian Crustal Rocks at Isua, West Greenland—Geochemical and Isotopic Evidence." *Earth and Planetary Science Letters* 27, no. 2: 229–39.

Moorbath, S., J. H. Allaart, D. Bridgwater, and V. R. McGregor. 1977. "Rb–Sr Ages of Early Archean Supracrustal Rocks and Amîtsoq Gneisses at Isua." *Nature* 270, no. 5632: 43–45.

Fedo, C. M., J. S. Myers, and P.W. Appel. 2001. "Depositional Setting and Paleogeographic Implications of Earth's Oldest Supracrustal Rocks, the > 3.7 Ga Isua Greenstone Belt, West Greenland." *Sedimentary Geology* 141: 61–77.

Furnes, H., M.J. de Wit, H. Staudigel, M. Rosing, and K. Muehlenbach. 2007. "A Vestige of Earth's Oldest Ophiolite." *Science* 315, no. 5819: 1704–07.

Furnes, H., M. Rosing, Y. Dilek, and M. J. de Wit. 2009. "Isua Supracrustal Belt (Greenland)—A Vestige of a 3.8 Ga Suprasubduction Zone Ophiolite, and the Implications for Archean Geology." *Lithos* 113, nos. 1–2: 115–32.

Komiya, T., S. Yamamoto, S. Aoki, Y. Sawaki, A. Ishikawa, T. Tashiro, et al. 2015. "Geology of the Eoarchean, > 3.95 Ga, Nulliak Supracrustal Rocks in the Saglek Block, Northern Labrador, Canada: The Oldest Geological Evidence for Plate Tectonics." *Tectonophysics* 662: 40–66.

Whitehouse, Martin J., Daniel J. Dunkley, Monika A. Kusiak, and Simon A. Wilde. 2019. "On the True Antiquity of Eoarchean Chemofossils–Assessing the Claim for Earth's Oldest Biogenic Graphite in the Saglek Block of Labrador." *Precambrian Research* 323: 70–81.

Vezinet, A., E. Thomassot, Y. Luo, D. G. Pearson, R. A. Stern, and C. Sarkar. 2022. "Zircon Geochronology and Hf–O Isotopes of the Nulliak Supracrustal Assemblage (Saglek Block–Canada): Constraints on Deposition Age and Setting, Metamorphic Age and Environments of Zircon Crystallization." *Precambrian Research* 379: 106789.

Sole, C., J. O'Neil, H. Rizo, J-L. Paquette, D. Benn, and J. Plakholm. 2025. "Evidence for Hadean Mafic Intrusions in the Nuvvuagittuq Greenstone Belt, Canada." *Science* 388, no. 6754: 1431–35.

A good overview of the geological origins of southern Africa can be found in:

Tankard, A. J., M. Martin, K. A. Eriksson, D. K. Hobday, D. R. Hunter, and W. E. L. Minter. 2012. *Crustal Evolution of Southern Africa: 3.8 Billion Years of Earth History*. Springer Science & Business Media.

I share Maarten de Wit's view that plate tectonics, much as we know it today, had started very early on, within a few hundred million years of Earth's origin. The following papers cover a range of ideas about the role of plate tectonics, or otherwise, on early Earth, focusing on the driving forces, vertical tectonics or different putative precursors to modern plate tectonics:

Nisbet, E. G., and C. M. R. Fowler. 1983. "Model for Archean Plate Tectonics." *Geology* 11, no. 7: 376–79.

England, P., and M. Bickle.1984. "Continental Thermal and Tectonic Regimes During the Archean." *Journal of Geology* 92, no. 4: 353–67.

Bickle, M. J. 1986. "Implications of Melting for Stabilisation of the Lithosphere and Heat Loss in the Archean." *Earth and Planetary Science Letters* 80, nos. 3–4: 314–24.

Hargraves, R. B. 1986. "Faster Spreading or Greater Ridge Length in the Archean?" *Geology* 14, no. 9: 750–52.

Van Kranendonk, M. J. 2011. "Cool Greenstone Drips and the Role of Partial Convective Overturn in Barberton Greenstone Belt Evolution." *Journal of African Earth Sciences* 60, no. 5: 346–52.

Moyen, J. F., and J. Van Hunen. 2012. "Short-Term Episodicity of Archean Plate Tectonics." *Geology* 40, no. 5: 451–54.

Korenaga, J. 2013. "Initiation and Evolution of Plate Tectonics on Earth: Theories and Observations." *Annual Review of Earth and Planetary Sciences* 41: 117–51.

Keller, B., and B. Schoene. 2018. "Plate Tectonics and Continental Basaltic Geochemistry Throughout Earth History." *Earth and Planetary Science Letters* 481: 290–304.

Stern, R. J. 2018. "The Evolution of Plate Tectonics." *Philosophical Transactions of the Royal Society A: Mathematical, Physical and Engineering Sciences* 376, no. 2132: 20170406. https://doi.org/10.1098/rsta.2017.0406.

Kusky, T. M., B. F. Windley, and A. Polat. 2018. "Geological Evidence for the Operation of Plate Tectonics Throughout the Archean: Records from Archean Paleo-Plate Boundaries." *Journal of Earth Science* 29, no. 6: 1291–1303.

Friend, C. R., and A. P. Nutman. 2019. "Tectono-Stratigraphic Terranes in Archean Gneiss Complexes as Evidence for Plate Tectonics: The Nuuk Region, Southern West Greenland." *Gondwana Research* 72: 213–37.

Johnson, T. E., C. L. Kirkland, N. J. Gardiner, M. Brown, R. H. Smithies, and M. Santosh. 2019. "Secular Change in TTG Compositions: Implications for the Evolution of Archean Geodynamics." *Earth and Planetary Science Letters* 505: 65–75.

Sobolev, S. V., and M. Brown. 2019. "Surface Erosion Events Controlled the Evolution of Plate Tectonics on Earth." *Nature* 570, no. 7759: 52–57.

Sobolev, A. V., E. V. Asafov, A. A. Gurenko, N. T. Arndt, V. G. Batanova, M. V. Portnyagin, et al. 2019. "Deep Hydrous Mantle Reservoir Provides Evidence for Crustal Recycling Before 3.3 Billion Years Ago." *Nature* 571, no. 7766: 555–59.

Kusky, T. M., J. Wang, L. Wang, B. Huang, W. Ning, D. Fu, et al. 2020. "Mélanges Through Time: Life Cycle of the World's Largest Archean Mélange Compared with Mesozoic and Paleozoic Subduction-Accretion-Collision Mélanges." *Earth-Science Reviews* 209: 103303. https://doi.org/10.1016/j.earscirev.2020.103303.

O'Neill, C., S. Marchi, W. Bottke, and R. Fu. 2020. "The Role of Impacts on Archean Tectonics." *Geology* 48, no. 2: 174–78.

Piccolo, A., B. J. Kaus, R. W. White, R. M. Palin, and G. S. Reuber. 2020. "Plume—Lid Interactions During the Archean and Implications for the Generation of Early Continental Terranes." *Gondwana Research* 88: 150–68.

Brown, M., T. Johnson, and N. J. Gardiner. 2020. "Plate Tectonics and the Archean Earth." *Annual Review of Earth and Planetary Sciences* 48: 291–320.

Korenaga, J. 2020. "Plate Tectonics and Surface Environment: Role of the Oceanic Upper Mantle." *Earth-Science Reviews* 205: 103185. https://doi.org/10.1016/j.earscirev.2020.

Zheng, Y. F., and G. Zhao. 2020. "Two Styles of Plate Tectonics in Earth's History." *Science Bulletin* 65, no. 4: 329–34.

Brenner, A. R., R. R. Fu, D. A. Evans, A. V. Smirnov, R. Trubko, and I. R. Rose. 2020. "Paleomagnetic Evidence for Modern-Like Plate Motion Velocities at 3.2 Ga." *Science Advances* 6, no. 17: eaaz8670. https://doi.org/10.1126/sciadv.aaz8670.

Korenaga, J. 2021. "Hadean Geodynamics and the Nature of Early Continental Crust." *Precambrian Research* 359: 106178. https://doi.org/10.1016/j.precamres.2021.

Kusky, T., B. F. Windley, A. Polat, L. Wang, W. Ning, and Y. Zhong. 2021. "Archean Dome-and-Basin Style Structures Form During Growth and Death of Intraoceanic and Continental Margin Arcs in Accretionary Orogens." *Earth-Science Reviews* 220: 103725. https://doi.org/10.1016/j.earscirev.2021.103725.

Palin, R. M., J. D. Moore, Z. Zhang, G. Huang, J. Wade, and B. Dyck. 2021. "Mafic Archean Continental Crust Prohibited Exhumation of Orogenic UHP Eclogite." *Geoscience Frontiers* 12, no. 5: 101225. https://doi.org/10.1016/j.earscirev.2021.103725.

Ning, W., T. Kusky, L. Wang, and B. Huang. 2022. "Archean Eclogite-Facies Oceanic Crust Indicates Modern-Style Plate Tectonics." *Proceedings of the National Academy of Sciences* 119, no. 15: e2117529119. https://doi.org/10.1073/pnas.2117529119.

Tarduno, J. A., R. D. Cottrell, R. K. Bono, N. Rayner, W. J. Davis, T. Zhou, et al. 2023. "Hadaean to PalaeoArchean Stagnant-Lid Tectonics Revealed by Zircon Magnetism." *Nature* 618, no. 7965: 531–36.

Fu, R. R., N. Drabon, B. P. Weiss, C. Borlina, and H. Kirkpatrick. 2024. "Statistical Reanalysis of Archean Zircon Paleointensities: No Evidence for Stagnant-Lid Tectonics." *Earth and Planetary Science Letters* 634: 118679.

Harrison, T. M. 2024. "We Don't Know When Plate Tectonics Began." *Journal of the Geological Society* 181, no. 4: jgs2023-212. https://doi.org/10.1144/jgs2023-212.

Lowe, D. R., and G. R. Byerly. 2024. "Geology of the Eastern Barberton Greenstone Belt, South Africa: Early Deformation and the Role of Large Meteor Impacts." *American Journal of Science* 324, no. 12: 2164–2182.

Index

Note: *Italic* page numbers refer to illustrations.

granodiorite, 40
gravimeters, 198
gravity force, and tectonic plate movement, 248
Great Dyke, Zimbabwe, 196
Great Marlborough Conglomerate, New
    Zealand, 137–39, *138*, 141; as birth of
    subduction zone, 140; comparison with Fig
    Tree Group, 141, 142; submarine landslides
    and debris avalanches, 138, 139, 140, 141
Great Oxidation Event, 272
greenhouse effect, 266, 267, 278, 279, 280
greenhouse gases, 27, 266, 267, 268, 273, 275,
    284; link with Milankovich cycles, 277–78;
    and ocean temperatures, 267
Greenland Ice Sheet, 47, 235, 277
"Green Like the Sky" (GLS), 26–27, 283,
    284–85
greenstone belts, 102, 252; and Archean
    mélange, 242–43; oldest, as fragments of
    oceanic crust or volcanic arcs, 247; sand
    grain distortion, 183–84; shapes and size,
    244–45; vertical rock layers, 157. *See also*
    Barberton Greenstone Belt; Belingwe
    Greenstone Belt, Zimbabwe
greenstones, 28, 37, 38, 40–41, 42, 84, 97;
    distinctive shapes, 42–43, 183–84; Kaapvaal
    Craton, 47; in other continents, 42
"gregarious batholiths," 38, 42–43, 58, 186
guerilla war, 09, 111, 186
gypsum, 271

Hadean Eon, 47, 235, 236, 238, 240, 243
Hadean rocks, 237
hafnium, 229, 230
Haggard, Rider: *King Solomon's Mines*, 195,
    210; *She: A History of Adventure*, 64
half-life, 43–44, 229, 233
Hall, Arthur Lewis: geological survey and map,
    33–34, 40–41; "Swaziland System," 34, 41,
    55, 99
Halley's Comet, 152
Halliday, Alex, 229
Hamersley Range, Western Australia, iron
    deposits, 205
heat engine, planetary, 239–40
heat flow, 89, 239
Heezem, Bruce, 130–31
Helen (British resident in Swaziland), 26–27, 29
helium, 225, 265
Herschel, John, 152; formation of new
    continents by "floating on a sea of lava,"
    152, 153; shape of the Earth, 152
Herschel, William, 152

Highveld of South Africa, 11, 12, 27, 31, 76, 183
Himalayas, 24–25, 66, 103, 185, 249, 281, 285
hippopotamus, 112
historical records, 285–87
Hlane Royal National Park, 257; waterhole and
    camp, 257–58
Hooggenoeg Formation of the Onverwacht
    Group, 58; komatiites and basalts, 103
horizontal squeezing: Barberton Greenstone
    Belt, 157; and mountain uplift, *156*
host country economy and mines, 199–200
Huangarua River, New Zealand, 165–69;
    angular unconformity, 166, 167, 169
Huangarua Syncline, 166, *167*, 168–69, 182;
    active growing in Aorangi Ranges, *170*
Hubble Space Telescope, 225
Huebeck, Chris, 182
Hunga volcano eruption, Tonga archipelago,
    224, 262; energy released, 262; lightning
    strikes and organic molecule synthesis,
    262
hunter-gatherers, 33
Hutton, James, 9
hydrocarbons, 270
hydrogen, 225, 265
hydrothermal activity, 196
hydrothermal vents, 266; fossilized
    microorganisms around, Onverwacht
    Group, 108; metals/minerals build up,
    197, 207, 208, 222; single-celled microbes
    around, 106, 107; as source of life, 262;
    West Mata volcano, "Shrimp City"
    communities, 92, 107

iceberg analogy of Earth's crust, 154, *154*,
    155, 157
ice cores, 277
igneous rocks, 7, 39, 80; categorization based
    on magnesium oxide content, 80–81;
    classification based on silica content, 80;
    dating, 44; silica content, 80
inner core, 227
interglacials, 276, 277, 285
international aid organisations operating in
    Swaziland, 189, 255
interstellar clouds, collapse, 225
iodine, 233
iron, 198, 200, 205–7, 229, 230; prevention
    from rusting due to lack of oxygen, 164,
    270–71; rusting, 273; rusting on Mars in
    oxygen-free atmosphere, 271; sources,
    206–7
iron meteorites, 226, 229

volcanic rocks, 7, 38, 42, 79–80; Barberton Greenstone Belt, *57*, 58; Fig Tree Group, 246; formation, 81. *See also* Onverwacht Group

Wairarapa, New Zealand, 164–66
warthogs, 149
water: early evidence on the Earth's surface, 237; in komatiites, 93, 246; as lubricant to weaken rocks, 250; on Mars and Venus, 268–69; in olivine crystals, 93; role in plate tectonics, 263–64
water vapor, 266, 268
weathering, 175–76, 178
weathering thermostat, 278–80
Wellman, Harold, 137
Westall, Frances, 108, 123
Western Australia: komatiites, 84; magnetism direction, 248; Pilbara Craton, 11, 47, 267; Shark Bay stromatolites, 111, 121
West Mata underwater volcano, 91–92, 262; boninite lava, 92, 93; hydrothermal vents ("Shrimp City"), 92, 107; pillow lavas, 92; subduction zone, 92
"which way up?", 51, 52, 53
wildebeest, 146, 149

Witwatersrand, South Africa, 183; annual gold production, 204; gold mines and mining, 201–4; gold rush, 201
written eyewitness accounts, 286–87

xenon, 233
x-ray fluorescence (XRF) spectrometer, 172–73

yellowcake, 216

Zagros Mountains, 180
Zealandia, 223
zebra, 149
zebra-like bands in gneiss, 38, 39
Zimbabwe, 108–9; author's eye injury, 112; Beitbridge border post managed by former Rhodesian Police, 109–10; Belingwe Greenstone Belt, 6, 111–12; gold mines and mining, *202*, 204–5; military vehicles escort convoys of civilian vehicles in, 110–11
Zimbabwe Craton, 38, 253; mineral deposits, 186
zinc, 100
zircon crystals, 46, 47, 48, 237; carbon detection in 4.1 billion-year-old, 237; magnetism preserved in, 243–44